Helmut Baumgärtel
Electrochemistry

Also of interest

X-ray Studies on Electrochemical Systems.
Synchrotron Methods for Energy Materials
Braun, 2017
ISBN 978-3-11-043750-8, e-ISBN 978-3-11-042788-2

Electrochemical Energy Systems.
Foundations, Energy Storage and Conversion
Braun, 2018
ISBN 978-3-11-056182-1, e-ISBN 978-3-11-056183-8

Bioelectrochemistry.
Design and Applications of Biomaterials
Cosnier (Ed.), 2019
ISBN 978-3-11-056898-1, e-ISBN 978-3-11-057052-6

Electrochemical Storage Materials.
From Crystallography to Manufacturing Technology
Meyer, Leisegang, Zschornak, Stöcker, 2018
ISBN 978-3-11-049137-1, e-ISBN 978-3-11-049398-6

Helmut Baumgärtel

Electrochemistry

A Guide for Newcomers

DE GRUYTER

Preface

This book has been written for those who are not familiar with electrochemistry but are interested in it. They may be engineers or physicists who use electrochemistry in their profession, And also teachers at high schools answering their pupils' questions concerning electrochemistry, like electromobility. This book is dedicated to those who want to know the story behind energy storage as reported in daily newspapers.

This book is not a textbook in the general sense. The author intends to conduct the reader from simple examples to ambitious problems that need a study in detail. Since electrochemistry is a complex topic, it requires understanding of the fundamental principles of electrode reactions and of the electric properties of electrode–surface interface.

https://doi.org/10.1515/9783110437393-202

Contents

Introduction

Electrochemistry is one of the branches of physical chemistry. The first pioneering observations were made at the end of the eighteenth century. During the nineteenth century the experiments of Volta, Faraday, Arrhenius and others were milestones in the development of electrochemistry. Later on, Nernst and his coworkers started a broad international development in this field. In 1896, Wilhelm Ostwald published the first textbook on electrochemistry.[1] Since the past century electrochemistry developed rapidly into different branches such as analytical chemistry, inorganic and organic chemistry, material science and medicine. Electrochemists have turned much of their attention to chemical problems. On one hand, sophisticated electronic instrumentation accompanied with the use of digital computers for running and controlling experiments and for data analysis and theoretical calculations considerably increased the experimental flexibility. Finally recent experimental developments in surface science influenced electrochemistry considerably. Methods like Scanning Tunnel Microscopy (STM) opened the microscopic view at solid electrodes in the nanoscopic scale.

This book has been written to describe the fundamentals, methods and applications in an up-to-date, understandable and systematic way. It is intended as a book for newcomers who will be introduced to electrochemistry, therefore the style is pedagogical throughout the book, and illustrations, graphs and tables have been presented to clarify the texts. I have tried to write it in a way that would make self-study possible by interested individuals.

The electrochemical cell is the experimental arrangement that is generally used in electrochemical research. It consists of at least two electrodes; there are different versions of electrochemical cells depending on the nature of electrodes involved. The energetics of the cell reaction can be described by thermodynamics.

The core of electrochemistry is the exchange of electric charge between an electrode and a substance, generally called "depolarizer," in the adjacent solution. This process occurs in a very thin layer at the boundary of the electrode by exchange of electrical charge (electrons or ions) across the interface of metal/solution. The structure, energetic and dynamic properties of this "double layer" are dominating the electrochemical events. Therefore a meaningful part of the following chapter "Fundamentals" is dedicated to the "double layer" (Chapter 2.4).

In the practical application of electrochemical processes and methods, however, the influence of the double layer is hardly directly observable, because the charge exchange across the phase boundary may be linked with other processes, like transport of depolarizer from the bulk of solution to the electrode surface,

1 Wilhelm Ostwald "Elektrochemie, ihre Geschichte und Lehre" Verlag von Veit & Comp. Leipzig 1896.

https://doi.org/10.1515/9783110437393-001

crystallization or formation of adsorbed layers at the surface of metal electrodes. In addition to these, chemical reactions in the bulk or at the surface may influence the manifestation of an electrode process.

The coupling of transport processes and/or chemical reactions to the charge exchange show a picture of electrochemistry, which seems at the first glance complicated and confusing for any newcomer to this field. It is one of the goals of this book to sort out the seeming confusions.

In order to reach this promising goal the text is organized into three main chapters: (1) Fundamentals, (2) Practical Work in Electrochemistry and (3) Applications in Electrochemistry. In each of these chapters the reader finds an overview of the most important topics. I hope the formulations are clear and instructive and not boring because of overcrowding details. To prevent this, in addition, "boxes" are added to the text. There is a twofold intension for these boxes: Some of them bring calculations illustrating formula given in the text, other contents are short compilations of related connections or data. It is necessary to study all the details for the reader who strives for deeper understanding and to look for the physical background. Such details are compiled in small chapters in the appendix.

Currently and in the near future, electrochemistry will play an irreplaceable role in modern techniques. This is illustrated in the part "Applications of electrochemistry." It consists of analytical application of electrochemical methods, the generation and storage of electrical energy, electrochemical synthesis, surface coating and corrosion. These topics, among others, will be adressed in the chapter "Applications of electrochemistry."

1 Units, constants and basic Laws

In this chapter the reader will become familiar with relevant quantities and with terms that are often used in electrochemistry.[2]

1.1 Energy

Energy can exist in different forms besides electrical energy: chemical energy, thermal energy (heat), energy of photons (light) and mechanical energy; these all can be converted into electrical energy.[3] Electrical energy plays a dominant role in context to electrochemistry; it is stored in different kinds of batteries. The transformation of chemical energy into electrical energy and the transformation of electrical energy into chemical energy in electrolysis are typical examples of electrochemical processes.

The primary units of energy employed are SI units (International System of Units); the base units of the SI system are kilogram for mass, meter for the measurement of length and second for time. In the older system of units, generally the cgs system was used, based on centimeter, gram and second.

The calorie is probably the most well-known unit of energy. It is defined by the energy necessary to heat 1 g (1 mL) of water from 14 °C to 15 °C at a normal pressure. In Table 1.1 different units of energy are compiled.

The density of energy is important for its storage. In a modern battery used in electromobiles 50 kWh may be stored; this corresponds to about 5 L of gasoline.

1.2 Power

Power is the flux of energy per unit of time (energy/time). The units of power are given by 1(J)/1(s) = 1 Watt (W) = 1 (Nm/s). Often used units are kilowatt (kW) or megawatt (MW) (1kW = 10^3 W or 1MW = 10^6 W)

2 Units and conventions are taken from Allen J. Bard, Roger Parsons, Joseph Jordan "Standard Potentials in aqueous solutions" Marcel Dekker Inc. New York Basel 1985. This is a rich compilation of electrochemical and thermodynamic data.

3 The energy content of the universe is constant.

https://doi.org/10.1515/9783110437393-002

Table 1.1: Different units of energy.

	(J)	(cal)	(eV)	(kWh)
1 Joule (J)	1	0.239	$0.624\ 10^{19}$	$2.78\ 10^{-7}$
1 calorie (cal)	4.185	1	$2.63\ 10^{19}$	$1.163\ 10^{-6}$
1 Electron-Volt (eV)	$1.602{\cdot}10^{-19}$	$3.38\ 10^{-20}$	1	$4.45\ 10^{-26}$
1 Kilowatt-Hour (kWh)	$3.6\ 10^{6}$	$0.86\ 10^{6}$	$2.25\ 10^{25}$	1
1 Terawatt.Year (TWa)	$3.16\ 10^{19}$			$8.77\ 10^{12}$
1 kg hard coal unit (SKE)	$2.93\ 10^{7}$	$7\ 10^{6}$		8.14

1 Joule (J) = 1 Watt-second (Ws) = 1 Newton-meter = 1 kg $m^2\ s^{-2}$ = 4.184 cal = $1.60\ 10^{19}$ eV.

1.3 Electrical units

1.3.1 Voltage and current

Volt is the unit of the electrical potential difference or is more well-known for the voltage between two points in an electronic device. It is defined as the SI base unit together with the Ampere A.

$$1\ \text{Volt} = 1\ \text{V} = 1\,\text{kg}\,\text{m}^2\,\text{s}^{-3}\,\text{A}^{-1} = 1\,\text{J}\,\text{A}^{-1}\,\text{s}^{-}.$$

Ampere is defined by the amount of electric charge passing through a cross-section of 1 cm^2 of a conductor per 1 s.

1.3.2 Electrical charge

In electrochemistry positively charged cations and negatively charged anions and electrons play an important role. The basic unit of electric charge is the Coulomb (C). It is defined by

$$1\,\text{Coulomb}\,(\text{C}) = 1\,\text{Amperesecond}\,(\text{As})$$

The elementary electrical charge e_0 can be determined as follows:

$$e_0 = F/N_a = 96494/6.0247\cdot 10^{23} = 1.60164\ \cdot\ 10^{-19}\,\text{C}$$

The elementary electrical charge e_0 ($1.602 \cdot 10^{-19}$ C) is not for practical use as it is not a functional unit; therefore, the Faraday constant F is mostly used as a unit of charge.

The Faraday constant F = 96484.456 C·mol^{-1} is given by the number of charged species present in 1 mol.

One mol is defined as the amount of substance containing N_A = $6.02\ 10^{23}$ atoms or molecules; N_A is the Avogadro's or Loschmidt's number.

Thus, 18 g water (1 mol) contains 6.02 10^{23} water molecules, and 65.36 g zinc contains 6.02 10^{23} zinc atoms.

The molecular weight is defined as the weight of one mol of molecules (atoms). This is related to an amount of substance containing $n_A = 6.02 \cdot 10^{23}$ molecules.

The combination *RT/F* is frequently used, which is expressed as follows:

$$RT/F = 0.02569\,\text{V or } (RT/F)\ \ln 10 = 0.059\,\text{V (at 25°C)}$$

R is the gas constant; $R = 8.314$ j mol^{-1} K^{-}

1.4 Ohm's law

Ohm's law most probably the readers know from their school days.

The basic Ohm's law connects the voltage *V* with the current *I*. It states that the voltage *V* at the end of a conductor is proportional to the current *I* in the conductor, that is, the ratio *V*/*I* for a given conductor has a constant value R, which is called the resistance of the conductor. Ohm's law is expressed as follows:

$$V = I \cdot R \tag{1.1}$$

The dimension of the electrical resistance *R* is given by the dimensions of the voltage *V* and the current *I*. If *V* is measured in absolute Volt and the current *I* in absolute Ampere, the resistance *R* results in absolute Ohm.

The electrical resistance *R* of a homogenous conductor is proportional to the length *L* and inversely proportional to the cross section *q*, expressed as follows:

$$R = \sigma\,(l/q) \tag{1.2}$$

The proportionality σ is defined as the "specific resistance" of the conductor. According to eq. (1.2) the dimension of σ is Ohm cm^{-1}; it defines the resistance of a conductor of length 1cm and cross section 1 cm^2.

The reciprocal value $1/\sigma = \kappa$ is called the "specific conductivity" of the conductor; the dimension of κ is Ohm^{-1} cm^{-1}, the numerical value is equal to the length (cm) of a conductor with a cross section of 1 cm^2, which has the resistance of 1 Ohm.

The Ohm's law is valid for homogeneous conductors and direct current (DC). For alternating current the relationship between voltage and current is more complex.

1.5 Faraday's laws

Michael Faraday is one of the earliest founders of electrochemistry. In 1833, he described two laws, which are known today by his name as Faraday's first law and Faraday's second law. These are quantitative laws expressing magnitudes of electrolytic effects.

The first law states that the amount of chemical change produced by current in an electrolysis is proportional to the quantity q of the transmitted electricity.

Faraday used the wiring diagram (Box 1, Fig. a) producing gas (H_2, O_2) by electrolysis of aqueous sulfuric acid, and measured the gas volume produced in electrolytic cells. He observed: $V_I = V_{II} + V_{III}$.

If during the contact time t at an electrode, m [mg] ions are deposited then these are m/M moles corresponding to (m/M) n N_A electrons. For the amount of charge q transported in the electrolyte results in the following expression:

$$q = (m/M) \cdot n \cdot N_A \cdot e_0 \tag{1.3}$$

$N_A = 6.0247 \cdot 10^{23}$ is the Loschmidt's number or Avogadro's number, which gives the number of species in 1 mol. The elementary charge of the electron is $e_0 = 1.60164$ 10^{-19} C.

The Faraday constant $F = 96487.0 \pm 1.6$ Coulomb is given by the following equation:

$$F = N_A \cdot e_0 \tag{1.4}$$

It corresponds to the amount of electrical charge of one mole of electrons or univalent ions.

The second law states: The amount of chemical changes produced by the same quantity of electricity in different substances (in solution) is proportional to their equivalent weights. In electrolytic reactions, the equivalent weight of a substance is the gram formula weight associated with the gain or loss of one electron.

The quotient of masses m_1/m_2 of two species 1 and 2, which are deposited at an electrode by the identical amount of charge q correspond to the quotient of their equivalent weights M_1/z_1 and M_2/z_2.

$$m_1/m_2 = (M_1/z_1)/(M_2/z_2) \tag{1.5}$$

Faraday discovered this law with the experiment device shown in Box 1 (Fig. a and b).

The electrolysis cells I–IV are connected in line. They are filled with different solutions, in cell I aqueous HCl is present and in cell II aqueous sulfuric acid is present, in cell III Cu^{2+} and in cell IV Ag^+ solution is present. After conduction of an identical current through the device the ratio of the masses of deposited material was I:II:III:IV = 1: 1: 31,8: 107,9; that is, the deposition follows the ratio of the equivalent weights.

The equivalent weight M/z is the ion weight M divided by the electrochemical valency z. Equivalents are H^+, Cl^-, Ag^+, ½ Cu^{2+}, 1/3 Fe^{3+}, ¼ $Fe(CN)_6^{4-}$ and so on. One gram equivalent of an univalent ion contains $N_A = 6.0247 \cdot 10^{23}$ ions (N_A = Avogadro's number). From F and N_A the electrical charge e_0 of a single ion can be calculated.

The Faradaic laws are the basis of coulometry (see Box 1).

With 1 C, 1.1180 mg of Ag are deposited, therefore for the deposition of 1 g atom of silver (107.88 g) one needs 107.88/0.001180 = 96494 Coulombs = 1 Faraday (F).

Box 1
Faraday's laws

It was Michael Faraday who in 1833 performed the following experiments:

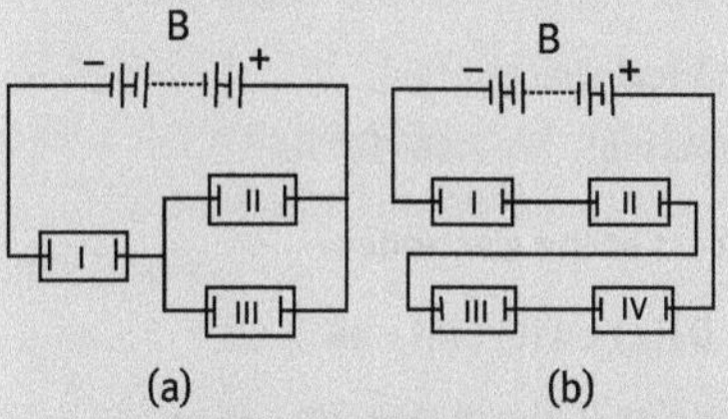

a) The current passes through three electrolytic vessels 1,2,3. Vessels 2 and 3 are turned parallel and both in series with vessel 1. The electrolyte is diluted sulfuric acid from which the electrolytic oxygen and hydrogen gases are produced. The observed gas volume from vessel 1 is the same as from vessels 2 and 3 together. The "First Law" of Faraday states that:

"The amount of deposited material is proportional to the transmitted electric charge."

$$m \approx I \cdot t$$

Michael Faraday found a second law.

b) Four vessels 1,2,3 and 4 are connected in line. They contain different electrolytes: vessel 1 with aqueous hydrochloric acid, vessel 2 with a solution of $CuCl_2$, vessel 3 with a solution of Na_2SO_4, and vessel 4 with a solution of $AgNO_3$. The ratio of deposited masses of hydrogen to deposited copper and silver is 1: 31.8: 107.9. This is exactly the ratio of the molar masses divided by the charge of the ions.

The second law of Faraday reads as follows:

"The ratio of masses of different substances deposited by the same amount of electricity corresponds with the ratio of their equivalent weights.

$$m_1/m_2 = (M_1/z_1)/(M_2/m_2)$$

The equivalent weight of an ion is the ion weight divided by the electrochemical valence. Equivalents are H^+, Cl^-, Ag^+, ½ Cu^{2+}, 1/3 Fe^{3+}, ¼ $Fe(CN)_6{}^{4-}$ etc.

For the conversion of one equivalent weight of material 96,490 Coulombs (A s = 1 Faraday) are transmitted into the electrochemical cell. The Faraday's laws are the base of coulometry, which is an important electroanalytical tool.

Utilizing the Faraday's laws

Problem

An electrical circuit contains an oxygen/hydrogen coulometer in line ("Knallgascoulometer") and a silver coulometer. Calculate the volume of the deposited oxygen/hydrogen ("Knallgas") at 0.960 bar and 298 K, if 856 mg silver are deposited in the silver coulometer.

Solution

In the silver coulometer, solvated silver ions are reduced to solid silver in the following reaction:

2 Fundamentals

This chapter describes some fundamental laws and models used in electrochemistry. The underlying principles of thermodynamics, interfacial structure, electrode kinetics, mass transfer and so on are also discussed, which are necessary for understanding the different experimental versions and the applications of electrochemistry. Appendices provide details of the underlying physics and detailed derivations for interested readers.

2.1 Electrochemical cells

2.1.1 A simple experiment

Figure 2.1 shows an example of a simple reaction. We put a rod of zinc or iron into an aqueous solution of copper sulfate. After a short time, we observe that copper metal gets deposited on the zinc rod, that is, the reduction of copper ions to copper occurs. Since the system remains electroneutral, there is clear evidence that the same amount of zinc atoms are oxidized to zinc ions, which are then dissolved in the solution.

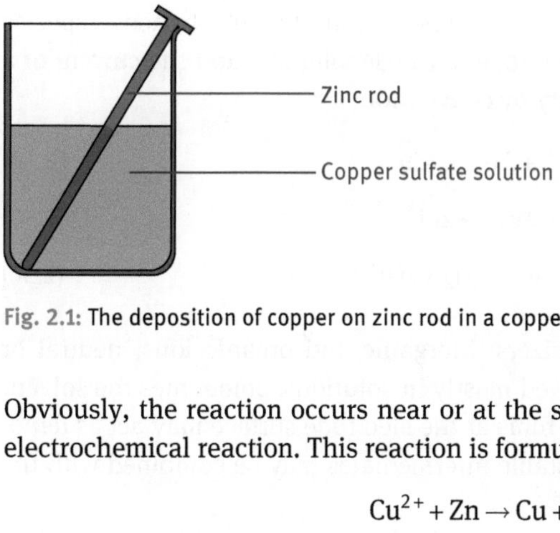

Fig. 2.1: The deposition of copper on zinc rod in a copper sulfate solution.

Obviously, the reaction occurs near or at the surface of the zinc rod. It is a typical electrochemical reaction. This reaction is formulated as follows:

$$Cu^{2+} + Zn \rightarrow Cu + Zn^{2+} \tag{2.1}$$

Sulfate ions that compensate the positive charge of the metal ions in the solution are not involved in the reaction.

The exergonic reaction (2.1) releases 212 kJ/mol of energy. However, we cannot use this energy as it is distributed in the solution in the form of heat. The reaction occurs voluntarily because the binding energy of electrons in copper is higher than in zinc.

https://doi.org/10.1515/9783110437393-003

How can we use the reaction energy?

The answer gives a simple electrochemical cell, which will be described in the forthcoming sections.

2.1.2 General aspects of the electrochemical cell

What is going on in the above-shown experiment?

To get better insight we separate the reaction partners into two compartments, thereby establishing an electrochemical cell. It is the central experimental device of electrochemistry. It contains at least two electrodes immersed in the solution of an electrolyte. At the anode, electrons are transferred from the depolarizer to the electrode, which means the depolarizer, is oxidized, for example,

$$Zn \rightarrow Zn^{2+} + 2\,e^- \tag{2.2a}$$

$$2\,Cl^- \rightarrow Cl_2 + 2\,e^- \tag{2.2b}$$

$$2\,H_2O \rightarrow O_2 + 4\,H^+ + 4\,e^- \tag{2.2c}$$

The term anode is exclusively defined by the oxidation of the depolarizer. The current of an anodic process is a positive quantity by convention.

Conversely, the cathode is defined by the reduction of an electroactive species by transferring electrons from the electrode to the depolarizer, and the current of a cathodic process is a negative quantity by convention:

$$Cu^{2+} + 2\,e^- \rightarrow Cu \tag{2.3a}$$

$$Cl_2 + 2\,e^- \rightarrow 2Cl^- \tag{2.3b}$$

$$2\,H_2O + 2\,e^- \rightarrow H_2 + 2OH^- \tag{2.3c}$$

There are different kinds of depolarizers: inorganic and organic ions, neutral or charged molecules, which are dissolved mostly in solutions. Sometimes the solvent itself, the electrode material itself or films at the electrode surface may act as depolarizers. In complex reactions, metastable intermediates may be combined with the electron transfer.

Two metal electrodes – an anode and a cathode – in contact with an electrolyte establish the simplest electrochemical cell.

To maintain electroneutrality in the electrochemical cell, the amount of reduction at the cathode and the amount of oxidation at the anode must be equal.

In addition to the equation of the anodic and the cathodic process, one gets the equation of the “overall cell reaction” from eqs. (2.2a) and (2.3a) as follows:

$$Zn + Cu^{2+} \rightarrow Zn^{2+} + Cu \tag{2.4}$$

In this equation, the number of transmitted electrons vanishes. It has the form of a conventional chemical reaction, which points to the possibility of applying the principles of chemical thermodynamics for describing the energetics of electrochemical reactions (see later).

Some selected electrochemical cells will be presented and described in the following sections.

2.1.3 The Daniell element

We will describe the Daniell element in some detail, because it is the prototype of an electrochemical cell. It consists of a zinc electrode and a copper electrode, both dipping into an aqueous solution of $ZnSO_4$ and $CuSO_4$, separated by a diaphragm, which hinders mixing of the electrolytes but is permissible for single ions. A schematic of the cell is shown in Fig. 2.2.

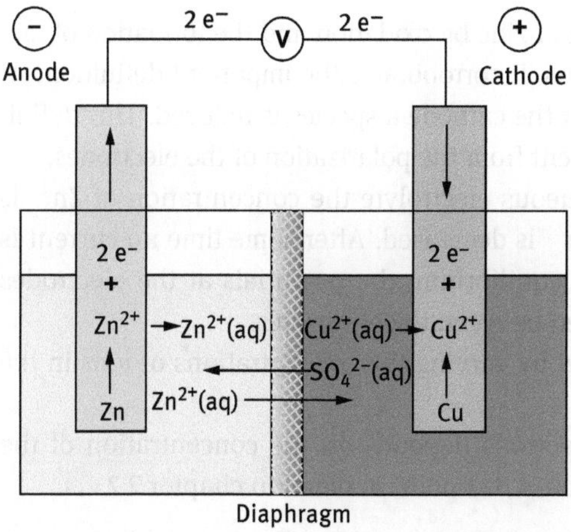

Fig. 2.2: A schematic representation of a Daniell element. The diaphragm separates the anodic space from the cathodic space; *V* is a voltmeter with an internal $R > 10^{14}\ \Omega$.

We connect the electrodes and observe a voltage. Owing to the high internal resistance of the voltmeter ($R > 10^{14}\ \Omega$), there is no current flow between the electrodes. The observed voltage gives evidence that the charged particles in both phases have different potential energies. Thus, the electrochemical cell is not in an equilibrium state.

The next step of our experimental excursion is to connect the electrodes via a low resistance line, which contains an ampere meter and a variable resistance.

Now there is a current flow between the electrodes. With decreasing resistance the current increases and the voltage V decreases.

The electrochemical cell is a closed circuit with an internal resistance. The voltage drop during current flow follows the Ohm's law:

$$U = R \cdot I \tag{2.5}$$

where U represents the voltage (V), R the resistance (Ω) and I the current (amp).

In an electrochemical cell, two different connections exist between the metal electrodes. On the one hand, the electrodes are connected by a metallic conductor, which hardly contributes to the internal resistance because the specific resistance of metals is very low. In this connection, the electrical charge is transported by electrons. On the other hand, the electrolyte connects the metal electrodes. It has a much higher resistance and is the decisive contribution to the internal resistance. Here ions transport the electrical charge.

In the Daniell cell, two electrode reactions occur:

$$Zn \rightarrow Zn^{2+} + 2e^- \tag{2.6a}$$

$$Cu^{2+} + 2e^- \rightarrow Cu \tag{2.6b}$$

The current arises from the dissolution of zinc by oxidation and the formation of metallic copper by reduction of copper ions. It corroborates the important definition: At the anode, a species is oxidized, and at the cathode a species is reduced. This definition of anode and cathode is independent from the polarization of the electrodes.

By the running current in the aqueous electrolyte the concentration of Zn^{2+} is increased and the concentration of Cu^{2+} is decreased. After some time no current is observed, which means the cell is in equilibrium, the potentials at the electrodes are identical and no voltage is observed between the electrodes.

The state of equilibrium is reached by varying the concentrations of ions in the electrolyte.

Obviously, the potential of an electrode depends on the concentration of the corresponding electrolyte. We will discuss this point in detail in chapter 2.2.

Now we start another experiment. We connect an outer high resistance voltage source to the cell (fig. 2.3); the voltage can be modified with the resistance R_1. We compensate the voltage between Zn and Cu electrodes until the current between the electrodes is zero. Now the cell is in equilibrium and the voltmeter shows 1.1 V. The value of the compensating tension is E_e, the equilibrium potential or rest potential of the cell. It is the difference of the two single potentials $E_{Cu/Cu^{2+}}$ and $E_{Zn/Zn^{2+}}$. In contrast to the rest potential, the potential of a single electrode cannot be measured.

As mentioned earlier, the potential of the electrodes depends on the concentration of the electrolyte. Thus, for the comparison of the equilibrium potentials of different cells the concentration of the electrolyte has to be standardized. The standard concentration is the activity $a = 1$. The activity takes into account the concentration-dependent intermolecular interactions of solvated ions, which is defined as

$$a = f\,c \tag{2.7}$$

where f is the activity coefficient. The value $a = 1$ is close to the concentration $c = 1$ (mol/L); a detailed discussion of ionic solutions is given later (chapter 2.3).

As mentioned earlier, the potential of a single electrode cannot be measured. Therefore, for the comparison of single electrode potentials one uses a standardized reference electrode. The internationally accepted reference electrode is the normal hydrogen electrode (NHE). It consists of a platinum electrode, which is dipped into an acid with the proton activity 1. The electrode is rinsed by gaseous H_2 at a pressure of 1 atm, and the temperature of the electrode is 25 °C.

The electrode reaction of the NHE is

$$H_2 \rightarrow 2\,H^+ + 2\,e^- \tag{2.8}$$

Walter Nernst introduced this electrode as the standard reference electrode and the electrochemical community agreed. It has arbitrarily been given the equilibrium potential 0 V and it is the origin of the electrochemical potential scale. The list of potentials of single electrodes refers to this arbitrary reference potential, and the real potential of the NHE will be discussed later.

We are now able to compare the potentials of the half-cells of the Daniell element versus the NHE:

$$Zn/Zn^{2+}\,//\,H^+/H_2 \qquad E_0 = -\,0.7628\,V$$
$$Cu/Cu^{2+}//\,H^+/H_2 \qquad E_0 = +\,0.3402\,V$$

It is evident that the standard equilibrium potential of the Daniell element is 1.103 V, which is the difference of the half-cell potentials. The zinc electrode is the negative pole and the copper electrode is the positive pole.

The overall cell reaction

$$Zn + Cu^{2+} \rightarrow Zn^{2+} + Cu \tag{2.9}$$

is a normal exergonic chemical reaction except that the electrons are not exchanged directly but through the metallic connection between the electrodes. The free enthalpy ΔG of the reaction is given by

$$\Delta G = n \cdot F \cdot \Delta\varphi \tag{2.10}$$

where n is the number of electrons exchanged per formula turnover, F is the Faraday constant, φ is the potential of a single electrode and $\Delta\varphi$ is the potential difference [V] between the anode and the cathode in the current-free cell, that is, the rest potential.

The electrochemical cell offers the possibility to control the overall cell reaction. Using the battery B (Fig. 2.3), the outer voltage source allows to influence the Daniell element. We can not only compensate the observed voltage until no current flows between the electrodes, if we overcompensate the potential difference at $U > E_e$, the electrons flow from the copper electrode to the zinc electrode. This switching of the direction of the overall cell reaction is a “privilege” of electrochemistry.

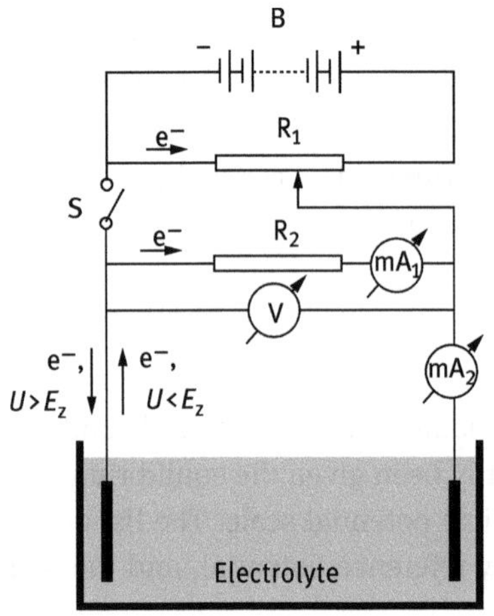

Fig. 2.3: An electrochemical cell is connected to the battery B. The voltage of the battery can be modified by the resistance R_1. V is a voltmeter, mA is an ampere meter. R_2 symbolizes the “inner resistance” of the cell.

This “privilege” is of outstanding interest in the frame of energy storage. The “self-driven” cell is a galvanic cell that converts chemical energy into electrical energy, and the “external-driven cell” is an electrolysis cell which stores electrical energy in the form of chemical energy. For more details, see Chapter 4.

2.1.4 More electrochemical cells

Now we discuss some other examples of electrochemical cells. The experimental setup is analogous to that shown in the previous example. The cell contains two platinum electrodes, the anode A and the cathode C. Between these electrodes exist two connections: the first one has a high-impedance voltmeter V ($\approx 10^{14}\ \Omega$) for measuring the potential difference between the electrodes and the second one is a low-impedance connection with an ampere meter to measure the current flowing between A and C. In addition to further investigations the electrodes are connected to the plus and minus poles of a battery B. By the variable resistance R, the voltage at the electrodes can be modified. The switch S allows the separation of the battery B from the electrochemical cell. We perform some simple experiments with this equipment.

At first the electrolyte in the cell is pure water. We observe no voltage at the voltmeter V, which means the states of both the electrodes are identical.

We put a voltage from the battery B at the electrodes A and C. At the ampere meter we observe only a very low current because the electrical conductivity of pure water is very low.

We continue the experiment with the same electrodes and with the same voltage, but the electrolyte is now diluted aqueous hydrochloric acid.

We observe an enhanced current at the electrodes, chlorine is deposited at the anode A and hydrogen at the cathode C.

The neutral solution contains solvated cations (H^+) and solvated anions (Cl^-). The formation of H_2 and Cl_2 indicates chemical reactions at the electrodes. We have an electrolytic cell.

The cations H^+ take up electrons, and are reduced at the cathode. The anions Cl^- are discharged at the anode:

$$2\,H^+ + 2\,e^- \rightarrow H_2 \quad (\text{cathode C}) \tag{2.11a}$$

$$2\,Cl^- \rightarrow Cl_2 + 2\,e^- \quad (\text{anode A}) \tag{2.11b}$$

The electrode C is the negative pole and the electrode A is the positive pole. Based on the polarization of the cell, the protons in the solution migrate to cathode C, and the chloride ions migrate to anode A. The electrons move In the outer circuit from the electrode A to the electrode C.

The addition of the electrode reactions leads to the overall cell reaction:

$$2\,H^+ + 2\,Cl^- \rightarrow H_2 + Cl_2 \tag{2.12}$$

The number n_e of transferred electrons per formula conversion is shown in eqs. (2.11a) and (2.11b) but not in eq. (2.12).

We have an electrolytic cell, which is driven by the applied potential. Between the electrodes A and C exists a potential gradient by which the electrons are “pumped“ from electrode A to the electrode C. Electrons move from the place with lower electrical potential to the place with higher electrical potential.

Obviously the ions are discharged at the surface of the electrodes by electrons that pass the low-resistance connection and cross the phase boundary metal/electrolyte. The electroactive species, protons and chloride ions, are called depolarizers. The educts (ions) are dissolved in the solution and the products of the electrode reaction (gases) escape to the atmosphere. Electrodes where only electrons cross the phase boundary are called redox electrodes.

We perform some additional experiments.

The electrolyte is now an aqueous solution of copper chloride ($CuCl_2$). We observe again a current. Like in the preceding experiment chlorine is developed at the anode A, at the cathode C the deposition of metallic copper is observed.

The electrode reactions are

$$Cu^{2+} + 2\,e^- \rightarrow Cu^0 \quad \text{(cathode C)} \tag{2.13a}$$

$$2\,Cl^- \rightarrow Cl_2 + 2\,e^- \quad \text{(anode A)} \tag{2.13b}$$

$$Cu^{2+} + 2\,Cl^- \rightarrow Cu^0 + Cl_2\ (n_e = 2)\ \text{overall cell reaction} \tag{2.14}$$

Electrodes where metal ions cross the double layer and are deposited at the electrode surface are called metal ion electrodes. At the surface the exchange of electrons occurs.

For the next experiment we substitute again the electrolyte. We use a diluted aqueous solution of H_2SO_4. We observe at the anode A the formation of oxygen O_2 and at the cathode C the formation of hydrogen H_2. The ratio of the volumes of H_2:O_2 is 2:1.

This is a rather complex reaction, because there are plenty of protons in the solution, whereas the concentration of OH^- ions in the acidic solution is extremely low.

The electrode reactions are

$$4\,H^+ + 4\,e^- \rightarrow 2\,H_2 \quad \text{(cathode C)} \tag{2.15a}$$

$$2\,SO_4{}^{2-} \rightarrow 2\,SO_4{}^{\bullet -} + 4\,e^- \quad \text{(anode A)} \tag{2.15b}$$

At the anode, sulfate radicals are formed. They are not stable in aqueous medium and undergo a chemical reaction with water molecules:

$$2\,SO_4{}^{\bullet} + 2\,H_2O \rightarrow 2\,H_2SO_4 + O_2 \uparrow \ \text{(chemical reaction)} \tag{2.16}$$

By summing up eqs. (2.15a), (2.15b) and (2.16), the following overall electrode reaction occurs:

$$2\,H_2O \rightarrow 2\,H_2 + O_2 \tag{2.17}$$

We learn from this example that the equation of the overall reaction describes not the mechanism of the electrode process and further on the final product of the electrolysis results by coupling of a primary electrode process with a chemical reaction. The evaluation of the mechanism of an electrode process can be difficult and time consuming.

There are other possibilities to modify the electrochemical cell. For instance, we use for the anode A silver instead of platinum and the electrolyte is now an aqueous solution of silver nitrate. We observe a current, but only weighting of the electrodes reveals that the silver anode lost material and the platinum cathode gained the same amount of material. The electrode process is given by the transfer of silver ions from the silver anode to the platinum cathode:

$$Ag^0 \rightarrow Ag^+ + e^- \quad \text{(anode A)} \tag{2.18a}$$

$$Ag^+ + e^- \rightarrow Ag^0 \qquad \text{(cathode C)} \tag{2.18b}$$

$$Ag^0{}_{\text{anode}} \rightarrow Ag^0{}_{\text{cathode}} \ \text{(overall cell reaction, } n_e = 1) \tag{2.19}$$

All these examples show that the equation of the overall cell reaction describes a chemical reaction; therefore, the laws of chemical thermodynamics can be applied for the calculation of the energetics of the reaction (see later).

2.1.5 The lead accumulator

The final example of an electrochemical cell, which will be described now, is the lead accumulator, known as the "starter battery" for driving the electric motor by which the internal combustion engine of a car is prepared to fire.

Imagine, you yourself want to realize such a lead accumulator. What are you going to buy? Of course you need a suitable vessel, then for the electrolyte diluted aqueous sulfuric acid (≈25%) and the electrodes, two pieces of very pure lead.

You dip one of the lead plates into the sulfuric acid and observe that the surface of the lead will be covered by a white layer. The potential $Pb/Pb^{2+} = -0.126$ V versus NHE; therefore, lead is oxidized to Pb^{2+} and protons are reduced to molecular hydrogen. Lead sulfate is hardly soluble in aqueous sulfuric acid. It covers the lead metal and prohibits further dissolution of the metal. Putting in a second plate of lead into the aqueous sulfuric acid, we get two identical electrodes. No voltage is observed between these electrodes.

Of course we prepared a very simple version of the lead accumulator, but it is suitable for further experiments.

We put about 20 V at the electrodes and we observe that the white coverage of the electrodes is substituted by a brown coverage at one of the electrodes, and the other electrode gets a metallic glance at the surface. After a while, gas evolution at the electrodes is observed, oxygen at the brownish electrode and hydrogen at the metallic electrode. Obviously, the electrolyte is decomposed in analogy to the above-mentioned experiment with sulfuric acid.

When we disconnect the electrodes from the external voltage source and measure the rest potential of the cell, it is 2 V. By connecting the electrodes via the ampere meter we observe a current. The current decreases in time and finally no current can be observed but we can observe again that both the electrodes are covered with a white layer. The lead accumulator is a rechargeable electrochemical cell. We observed one cycle of loading the cell in the driven electrolytic state and deloading it in the galvanic state of the cell.

The electrode reactions are

$$PbSO_4 + 2e^- \rightarrow Pb^0 + SO_4{}^{2-} \qquad \text{(cathode)} \tag{2.20a}$$

2.2 Electrochemistry and thermodynamics[4]

We know that the difference between the initial state of an electrochemical cell and its final state can be described by a chemical reaction, for example, in the Daniel element by equation 2.24.

$$Zn + Cu^{2+} \longleftrightarrow Zn^{2+} + Cu \tag{2.24}$$

Between the initial state and the final state exists an energy difference ΔG. This energy difference ΔG is the maximum net energy which can be gained by transferring electric charge from an initial potential to a final potential. It is defined by the difference of enthalpy H and entropy S:

$$G = H - \mathrm{T\,S} \tag{2.25}$$

or

$$G_{\mathrm{initial}} - G_{\mathrm{final}} = \Delta G = \Delta H - T\Delta S \tag{2.26}$$

Now we need the thermodynamic relation between the equilibrium cell voltage ε_0 and the free enthalpy ΔG.

If a differentially small current flows in the galvanic cell, which can be realized by a counterpotential (battery) or by applying an adequate large resistance in the circuit of current, the cell reaction occurs very close to the equilibrium state with a turnover in accordance with Faraday's law.

To be clear we repeat the single electrode processes in the cell:

$$Cu/Cu^{2+}(aq)/Zn^{2+}(aq)/Zn \tag{2.27}$$

Reaction (2.24) results from the addition of the single electrode processes (2.28a) and (2.28b):

$$Cu \longrightarrow Cu^{2+} + 2\,e^- \ \text{(anode)} \tag{2.28a}$$

$$Zn^{2+} + 2\,e^- \longrightarrow Zn \quad \text{(cathode)} \tag{2.28b}$$

For deriving the relationship between cell voltage and free enthalpy we chose the course of the reaction arbitrarily. We consider according to reaction (2.24) a complete formula turnover and we know that two electrons (n = 2) are transferred. The amount of transferred electricity is

$$Q = I \cdot t = n \cdot F \tag{2.29}$$

where Q is the amount of charge that has been transferred from one electrode to the other electrode and

4 Additional details of thermodynamics are described in Appendix A.

$$Q\varepsilon = nF\varepsilon \tag{2.30}$$

is the amount of energy that has been transferred between the electrodes.

Here thermodynamics means equilibrium thermodynamics, which means it can be applied to the difference of equilibrium states.

This brings us in a difficult position because: How can we transfer charge at an equilibrium potential ε_0 where the current per definition is zero? We perform the charge transfer at a differential small deviation from the equilibrium potential where the current is unobservably small. Thus the charge transfer takes infinite time, and it cannot be realized experimentally. Nevertheless, the derived relations are valid, because the equilibrium thermodynamics contains not the time. The cell reaction is reversible in the thermodynamic view and we can use ε_0 instead of ε.

The transferred energy $n\ F\ \varepsilon$ taken from the environment enhances the energy of the cell if $n\ F\ \varepsilon$ is positive. This is the case when the cell is an electrolytic cell. If $n\ F\ \varepsilon$ is negative we have a galvanic cell, which delivers energy to the environment.

Thus we get

$$\Delta G = nF\varepsilon_0 \tag{2.31}$$

or

$$\varepsilon_0 = \Delta G/nF = \Delta G\ (\text{kcal/mol})/n23.06(\text{kcal/Vmol}) \tag{2.32}$$

ΔG depends on the following variables: pressure p, temperature T and the activity a_i (concentrations) of the i species involved in the reaction:

$$\Delta G = f(p,\ T,\ a_i) \tag{2.33}$$

Thermodynamics describes changes of G or ΔG by the total differential[5]

$$d\Delta G = (d\Delta G/dp)_{T,\,a_i} + (d\Delta G/dT)_{p,\,a_i} + \sum (d\Delta G/da_i)_{p,\,T} \tag{2.34}$$

Activity a_i is introduced by the chemical potential μ_i for each species involved in the cell reaction:

$$\mu_i = \mu_i{}^0 + RT\ \ln a_i \tag{2.35}$$

Then eq. (2.34) can be rewritten as

$$d\Delta G = (d\Delta G/dp)_{T,\,a_i} + (d\Delta G/dT)_{p,\,a_i} + \sum (d\Delta G/d\mu_i)_{p,\,T} \tag{2.36}$$

5 For the meaning of a total differential, see Appendix A.

Using eq. (2.31) we derive the dependence of ε_0 from pressure p, temperature T and concentration c_i, respectively, the activities a_i of the species i involved in the electrode reaction.

2.2.1 The temperature dependence of ε_0

The temperature dependence of ε_0 follows from eq. (2.36). Provided with dp is zero (isobar) and $d\mu_i = 0$ (equilibrium) we get the temperature dependence of the equilibrium potential ε_0

$$(d\varepsilon_0/dT)_{p,\,a_i} = 1/nF\,(d\Delta G/dT)_{p,\,a_i} \tag{2.37}$$

The temperature coefficient of the free enthalpy is the change in the entropy $-\Delta S$[6]:

$$(d\Delta G/dT)_{p,\,a_i} = -\,\Delta S \tag{2.38}$$

The temperature coefficient delivers an important relation between the heat formation and heat consumption, respectively, during the cell reactions. Details will be presented in Chapter 4.

2.2.2 The pressure dependence of ε_0

Thermodynamics tells us that the dependence of the free reaction enthalpy from the pressure for one formula turnover is given by

$$(d\Delta G/dp)_{T,\,a_i} = \Delta V \tag{2.39}$$

Introducing $\Delta G = n\,F\,\varepsilon_0$ results in eq. (2.40)

$$(d\,\varepsilon_0/dp)_{T,\,a_i} = RT/nF \cdot 1/p \tag{2.40}$$

By integration of this equation we get

$$\varepsilon_0 = \varepsilon + RT/nF\ \ln p \tag{2.41}$$

The pressure dependence of the potential is important if gas electrodes are involved in an electrochemical cell. A gas electrode consists of a stable metal, for example, platinum, which dips into the liquid electrolyte. In the electrolyte the reactive gas is dissolved under a defined partial pressure. At the surface of the electrode, a redox electrode, electrons are exchanged.

6 This relation is derived explicitly in the Appendix A.

The most important gas electrode is the hydrogen electrode, because the NHE (Normal Hydrogen Electrode) defines the origin of the electrochemical potential scale ($\varepsilon_0 = 0$).

The pressure dependence of this electrode can be controlled experimentally by establishing an electrochemical cell with a pressure-independent reference electrode, like the silver/silver chloride electrode:

$$Pt(H_2)/H_2, HCl\,(aq)\,//\,HCl(aq), AgCl/Ag \tag{2.42}$$

The brutto cell reaction is given by

$$H_2 + 2\,AgCl \longrightarrow 2\,HCl + 2\,Ag \tag{2.43}$$

where per formula turnover two electrons are consumed ($n = 2$).

The pressure dependence (eq. (2.40)) for this cell is given by

$$\varepsilon_0 = \varepsilon - RT/nF\,\ln p(H_2) \tag{2.44}$$

Due to the compression of hydrogen in the cell reaction we get the negative sign in eq. (2.44). The pressure dependence of the hydrogen electrode follows this equation in the range 10^{-2} to 10^3 atm.

2.2.3 The dependence of ε_0 on the concentrations a_i

The voltage ε_0 of an electrochemical cell depends on the concentrations c_i (more exactly activities a_i) of all the species i involved in the cell reaction. Under isothermal conditions ($dT = 0$) and isobar conditions ($dp = 0$), ε_0 depends exclusively on the concentrations of the involved species. We derive now one of the most important equations of electrochemistry because the mentioned boundary conditions are fulfilled in most of the electrochemical experiments.

For the illustration of the following derivation we use an electrochemical cell, which consists of an arbitrarily chosen redox electrode and the hydrogen electrode. The electrode reaction of these two electrodes is

$$(-\nu_1)\,S_1 + (-\nu_2)\,S_2 \longleftrightarrow (\nu_3)\,S_3 + (\nu_4)\,S_4 + n\,e^- \tag{2.45a}$$

$$nH^+ + ne^- \longleftrightarrow n/2H_2 \tag{2.45b}$$

(ν_i) represents the stoichiometric factor, which is negative at the side where species are consumed and positive at the right side where species are generated; i is the running number for the involved species. Summing up eqs. (2.45a) and (2.45b) results in the equation of the Brutto cell reaction.

Now we remember the relation between the free enthalpy and the potential $\boldsymbol{\varepsilon_0}$:

$$\Delta G = n\,F\,\varepsilon_0 \tag{2.46}$$

ΔG results from the sum of chemical potential μ_i of the involved species i.The chemical potential μ_i depends on the activities a_i of each of the species i. It is

$$\mu_i = \mu_i' + R\,T \ln a_i \tag{2.47}$$

Substitution of eq. (2.47) into

$$c = 1/nF \sum v_i \mu_i \tag{2.48}$$

yields the famous Nernst equation

$$\varepsilon_0 = E_0 + R\,T/nF \sum v_i \ln a_i \tag{2.49a}$$

or in the more popular form for a redox couple

$$\varepsilon_0 = E_0 + R\,T/nF \;\ln c_{Red}/c_{Ox} \tag{2.49b}$$

which combines the potential of an electrochemical cell with the activities (concentrations) of the species i involved in the cell reaction. At low concentrations $c < 10^{-3}$ mol/L activities can be substituted by concentrations, the activity coefficient f is close to 1 (see chapter 2.3.7).

Some comments on eq. (2.49) seem to be useful. E_0 is now the potential of the cell measured versus the NHE. Because the potential of the NHE is arbitrarily taken to be zero, E_0 can be assigned to the potential of the coupled half-cell. Of course, the values of E_0 published in many compilations of potentials of single electrodes presuppose that $R\,T/n\,F\,\Sigma v_i \ln a_i$ is zero, that is $a_i = 1$. Provided these conditions are fulfilled, E_0 is the normal potential of the half-cell, that is, the single electrode is coupled to NHE.[7]

The factor $R\,T/F$ is repeatedly used. It has the dimension of a potential and for $n = 1$ comes to 25.6 mV if the natural logarithm is used. In case you use the decadic logarithm this factor comes to 59.2 mV at 25 °C. It is one of the mostly used numbers in electrochemistry.

Box 2
Electrochemical thermodynamics

The two examples given in this box illustrate the relation between electrochemistry and thermodynamics.

1. Calculation of the free standard enthalpy of the reaction

$$Ag + 1/2\,Hg_2Cl_2 \rightleftharpoons AgCl + Hg$$

7 It must be emphasized that the potential of the NHE is not zero. This is an internal standardization. The real potential of the NHE is about 4.7 eV. But since potentials are used to compare the potential of different single electrodes the common procedure is useable.

We use the equation

$$\Delta G^0 = -nFE^0$$

We break down the reaction into two partial reactions

$$AgCl + e^- \longrightarrow Ag + Cl^- \quad E^0{}_1 = 0.2223\ V^*$$

$$1/2\, Hg_2Cl_2\, {}^* e^- \rightarrow Hg + Cl^- E^0{}_2 = 0.2682\ V^*$$

(*These values have been taken from tables on standard potentials.)

The difference $E^0{}_2 - E^0{}_1$ of the standard potentials yields the standard potential of the reaction under consideration

$$E^0 = E^0{}_2 - E^0{}_1 = 0.2682 - 0.2223 = 0.0459\,V$$

$$n = 1,\quad F = 96,485\ C/mol$$

$$\Delta G^0 = -1\,(96,485\ C/mol)\,(0.0459\,V) = -4.43\ kJ/mol$$

2. Calculation of an equilibrium constant

What is the equilibrium constant of the reaction

$$Fe^{2+} + Ce^{4+} \rightleftharpoons Fe^{3+} + Ce^{3+}$$

at a temperature of 298 K?

The equilibrium constant is obtained by using

$$K = \exp(-\Delta G^0/RT)$$

where the free standard enthalpy ΔG^0 is calculated from electrochemical data using

$$\Delta G^0 = -nFE^0$$

The reaction

$$Fe^{2+} + Ce^{4+} \rightleftharpoons Fe^{3+} + Ce^{3+}$$

is split into the partial reactions

$$Fe^{3+} + e^- \rightleftharpoons Fe^{2+} \quad E^0{}_1 = +\,0.770\,V^*$$

$$Ce^{4+} + e^- \rightleftharpoons Ce^{3+} \quad E^0{}_1 = +1.443\,V^*$$

(*These values have been taken from tables on standard potentials.)

We get

$$E^0 = E^0{}_2 - E^0{}_1 = 1.443\,V - 0.770\,V = 0.673\ V$$

$$z = 1,\ F = 96,485\ C/mol$$

$$\Delta G^0 = -1\,(96,485\ C/mol)\,(0.673\ V) = -\ 64.9\ kJ/mol$$

and K is calculated by

$$K = \exp(-64.9\ \cdot\ 10^3\,J/mol/(8.314\,J/mol\,K)\,(298\,K) = 2.4\ \cdot 10^{11}$$

l is the distance between the electrodes. The electrical field represents a force that accelerates electrically charged species (see Appendix B). This force F_{el} is given by

$$F_{el} = q\,|E| = z \cdot e \cdot U/l \tag{2.51}$$

where q is the charge of an ion.

The electrical field brings about the migration of the ions. Cations are accelerated in direction to the negative charged electrode and anions in direction to the positive charged electrode. However, the collision with solvent molecules cuts down the initially accelerated movement of ions within about 10^{-13} s. Thus, a constant velocity v_i of the ions results. There exists an equilibrium between the electric force F_{el} and the frictional force F_f. We apply the Stokes law, which describes the frictional force F_f

$$F_f = -\,6\pi\eta\; r_i\, v_i \tag{2.52}$$

where η is the viscosity of the solvent, r_i is the radius of the ion and v_i is the velocity of the ion i.

The velocity v_i is constant when the relaxing force compensates the accelerating force:

$$F_e = F_f \tag{2.53}$$

In the frame of this simple model the ions are treated as spheres moving in a viscous fluid with an averaged drift velocity v_i. The averaged velocity v_i of the ions can be expressed by the following equation:

$$v_i = z_i\, e\, E/6\pi\,\eta\, r_i \tag{2.54}$$

Deriving the electric conductivity of the electrolyte the velocity v_i is of minor interest, because it depends not only on z_i and r_i but also on η and E. Therefore, the mobility u_i of the ions is defined:

$$u_i = v_i/E = z_i\, e/6\pi\,\eta\, r_i \tag{2.55}$$

The mobility is the velocity of the ion in a field of unit strength (1 V/m).

The ion mobility depends on the viscosity η of the solvent and via η it depends on the temperature and the pressure. Moreover, the ion mobility depends on the concentration of the ions and on the radius of the ions. This radius is not equivalent to the crystal radius determined by X-ray investigation of solids. The ions in the liquid are solvated and their radius depends on the number of neutral solvent molecules attached to the ion. In addition at higher concentrations the ion–ion interaction cannot be neglected. Therefore, it should be emphasized that our model is valid for diluted solutions.

We shall later see that typical values of the mobility lie in the range from $4 \cdot 10^{-8}$ to $9 \cdot 10^{-8}$ m²/sV, which corresponds to a drift velocity of $4–9 \cdot 10^{-3}$ mm/s. This seems to be slow but it is not when it is expressed on a molecular scale. It corresponds to an ion passing about 10,000 solvent molecules each second.

Now we need a relation between the measurable current I and the ion mobility u.

The mobility of the ions influences the current I flowing under a given voltage U through the electrolyte. This current can easily be measured in the external circuit. We need a relation between this current I and the mobility u, respectively, the velocity v_i of the ions.

We start with the definition of the current, that is, the number of electric charges crossing the unit cross-section per second:

$$I = q/t \tag{2.56}$$

We evaluate the relation between the current I and the velocity v of ion migration introducing some assumptions.

For the following consideration the electrolyte shall consist of only two kinds of ions and it is fully dissociated. This yields per formula unit ν^+ cations with the charge z^+ and ν^- anions with the charge z^-. In the conduction cell we define a distance l between the electrodes that have the area A. The area A of the electrodes is oriented perpendicular to the direction of the ion flux. The concentration c of the electrolyte is given by $c = n/V$ (mol/L). Between the electrodes exists the voltage U (V).

The current I crossing the electrolyte is given by the total charge q of the positive and the negative ions arriving at the planar electrodes of the cell during the time t.

During the time t ions travel over the average distance x in the direction of the electric field:

$$x = \mathrm{v} \cdot t \tag{2.57}$$

where v is the velocity of the ions and as per definition t is 1 s.

How much ions reach the electrodes per second?

All the cations that are separated from the cathode by the distance $\mathrm{v}^+ \cdot t$ and all the anions that are separated from the anode by the distance $\mathrm{v}^- \cdot t$ will reach the electrodes after 1 s. N ions will pass an imaginary window of the cross-section A during the time t:

$$N = c \cdot N_A \cdot A \cdot \mathrm{v} \cdot t \tag{2.58}$$

where c (mol/L) is the concentration of ions, $N_A = 6.022 \cdot 10^{23}$ (ions)/mol is the Avogadro number and v is the velocity of the ion.

The velocity of the anions and the cations may be different; therefore, eq. (2.58) splits into two contributions:

$$N^+ (\text{cations with the charge } z^+ \, e) = \mathrm{v}^+ \, c \, N_A \, A \, \nu^+ \, t \tag{2.59a}$$

$$N^- (\text{cations with the charge } z^- \, e) = \mathrm{v}^- \, c \, N_A \, A \, \nu^- \, t \tag{2.59b}$$

To avoid big numbers we introduce the Faraday constant $F = N_A \, e_0$. Adding up the two contributions we get the total current I:

$$I = q/t = F \cdot A\,(\mathrm{v}^+\, cz^+\, v^+ + \mathrm{v}^-\, cz^-\, v^-) \qquad (2.60)$$

The velocity of the ions depends on the field strength $E = U/l$; therefore, multiplying eq. (2.60) by E and taking into account eq. (2.55) $v_i = |E|u_i$ or $v_i = u_i\, U/l$ results:

$$I = (F \cdot A/l)\,(\mathrm{v}^+\, cz^+\, u^+ + \mathrm{v}^-\, cz^-\, u^-) \cdot U \qquad (2.61)$$

According to eq. (2.61) the current I is, as expected, proportional to the voltage U between the electrodes. This corresponds to the Ohm's law.

This result can be proofed experimentally.

As shown in Fig. 2.5, in using alternating current (AC) or direct current (DC) voltage the experiment reveals a discrepancy.

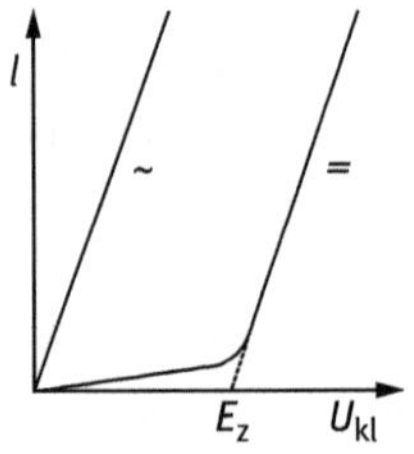

Fig. 2.5: Current–voltage curve of an electrolytic cell using DC voltage or AC voltage.

This figure shows the current–voltage curves of an electrolytic cell using DC or AC voltage. The DC current below the potential E_Z increases slowly (i.e., high cell resistance) and above the potential E_z the current increases strongly. The reason for this apparent discrepancy is a property of the phase boundary electrode/electrolyte, where an electrolytic double layer is formed and creates an additional resistance. Using AC voltage, the double layer is loaded and deloaded with the frequency of the AC voltage. The current is a linear function of the voltage. A second complication is that DC could lead to an electrolytic reaction at the electrodes and so the nature of the solution in their vicinity would change as the measurement proceeded.

If we use AC voltage at the electrodes we avoid these complications and find the expected proportionality between current and voltage in accordance with the Ohm's law.

2.3.3 Resistance and conductivity

How to measure the resistance of the electrolyte?

The classical device for measuring a resistance is the "Wheatstone bridge" (Fig. 2.6). It is established by four resistances, two of them are fixed as their values are known, one of them is a adjustable resistance and the fourth is the resistance of the electrolyte with an unknown value, which shall be determined.

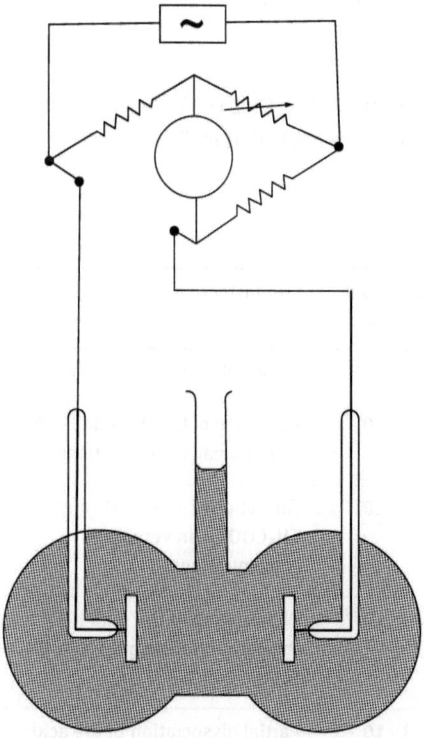

Fig. 2.6: Wheatstone bridge of resistance to measure the resistance of an electrolyte.

Two branches of the bridge are connected by a high impedance voltmeter. An AC source with about 1 kHz delivers a given voltage. With the tunable resistor the bridge is equilibrated, which can be seen at the voltmeter. For the equilibrated bridge holds the simple relation $R_1{:}R_2 = R_t{:}R_x$, where R_t is the value of the tuneable resistance and R_x is the value of the wanted resistance, that is, the resistance of the electrolyte.

Now we have measured the resistance R and we know the reciprocal $1/R$, the value of the conductivity.

In eq. (2.61) the factor $(F \cdot A/l)\ (v^+\ c\ z^+\ v^+ + v^-\ c\ z^-\ v^-)$ is the reciprocal resistance $1/R$:

$$1/R = F\,(v^+\,cz^+\,u^+ + v^-\,cz^-\,u^-)A/l \tag{2.62}$$

The resistance R still depends on the cross-section A and the distance l between the electrodes.

2.3.3.1 Specific resistance

For practical use it makes sense to define the specific resistance ρ

$$\rho = RA/l \tag{2.63}$$

or even better the reciprocal of ρ, the electric conductivity κ

$$\kappa = 1/\rho = F\,(c^+ \cdot z^+ \cdot u^+ + c^- \cdot z^- \cdot u^-) \tag{2.64}$$

The electric conductivity of an aqueous 1 M solution of NaCl is $0.74 \cdot 10^{-3}\ \Omega^{-1}\ \mathrm{cm}^{-1}$.

The electric resistances of different materials are compiled in Table 2.1.

Table 2.1: Conductivity of different solutions and materials.

System	T (°C)	Conductivity κ Ω^{-1} cm^{-1}	Conductivity by
Benzene	20	$5 \cdot 10^{-14}$	Dissociation of water impurities
Methanol	25	$2–7 \cdot 10^{-9}$	Formation of CH_3O^- and $CH_3OH_2^+$ in very small concentrations
Acetic acid (100%)	25	ca. $4 \cdot 10^{-9}$	Formation of <Ch^3COO^- and $CH_3COOH_2^+$ in very small concentrations
Water (very pure)	25	$6.48 \cdot 10^{-8}$	Dissociation in OH^- and H_3O^+ and the concentration of H^+ and OH^- is 10^{-7} mol/L
Aqueous acetic acid $c = 1$ mol/L	20	$1.3 \cdot 10^{-1}$	Partial dissociation of the acid into CH_3COO^- and H_3O^+
Aqueous LiCl $c = 1$ mol/L	18	$0.634 \cdot 10^{-1}$	Dissociation into Li^+ and Cl^-
LiCl in methanol $c = 1$ mol/L	20	$0.183 \cdot 10^{-1}$	Dissociation into Li^+ and Cl^-
Aqueous NaCl $c = 1$ mol/L	18	$0.744 \cdot 10^{-3}$	Dissociation into Na^+ and Cl^-
Aqueous $MgSO_4$ $c = 1$ mol/L	18	$0.438 \cdot 10^{-1}$	Dissociation into Mg^{2+} and SO_4^{2-}
Aqueous KOH $c = 1$ mol/L	18	$0.184 \cdot 10^{-1}$	Dissociation into K^+ and OH^-
Aqueous H_2SO_4 $c = 1$ mol/L	18	$0.366 \cdot 10^{-1}$	Dissociation into H_3O^+ and SO_4^{2-}
Ceramics, $ZrO_2 + Y_2O_3$	1,000	$0.5 \cdot 10^{-1}$	O_2^- as mobile charge carriers
NaCl melt	1,000	4.17	Complete dissociation into Na^+ and Cl^-
Mercury	0	$1.063 \cdot 10^4$	Electrons in the conduction band
Copper	0	$6.45s \cdot 10^5$	Electrons in the conduction band

The molar conductivity Λ

As noticed in eq. (2.62), the electric conductivity depends on the concentration c^+ and c^- of the cations and anions. Division by the concentration c gives the molar conductivity Λ of the electrolyte, which consists in the contribution of the cations and of the anions:

$$\Lambda^+ = F \cdot z^+ u^+ \tag{2.65a}$$

$$\Lambda^- = F \cdot z^- u^- \tag{2.65b}$$

The molar conductivity depends on the ion mobility u and on the charge number z^+ and z^- which indicates the amount of transported charge.

Thus the molar conductivity of the electrolyte can be written as

$$\Lambda = \nu^+ \cdot \Lambda^+ + \nu^- \cdot \Lambda^- \tag{2.66}$$

This is the first law of Kohlrausch. It postulates the independent migration of cations and anions.

How are specific conductivity and molar conductivity dependent on the concentration?

We used a very simple model to derive the electric conductivity and therefore the question arises: How the calculated values correspond with the experimental data?

From the equation

$$\kappa = 1/\rho = F \cdot c\,(\nu^+ z^+ u^+ + \nu^- z^- u^-) \tag{2.67}$$

follows that the conductivity should be proportional to the concentration of the electrolyte,

Figure 2.7 shows a diagram in which κ is plotted against the concentration c. At a first glance, there seems to be no satisfying correspondence between κ and c. However, some features of the diagram fit the expectation. The curves start at the origin and at low concentrations they are in good linear approximation. On the other hand, with increasing concentration there are considerable deviations from linearity. There appears a maximum and at higher concentrations the conductivity decreases with increasing concentration.

In Fig. 2.8, the molar conductivity is plotted versus low concentrations. According to eqs. (2.65a) and (2.65b), one expects that Λ is independent of the concentration. However, this is not the case.

The 1:1 electrolytes show only a small dependence on the concentration; however, the 1:2 electrolytes (H_2SO_4, $MgSO_4$) show considerable deviation from the expected behavior and the acetic acid is completely out of the expectation.

Kohlrausch found empirically the following relation, known as Kohlrausch's square root law:

$$\Lambda_c = \Lambda_0 - k\sqrt{c} \tag{2.68}$$

Provided the molar conductivities Λ_0^+ and Λ_0^- are known, the molar conductivity of a salt can be calculated as shown in the following example:

$\Lambda(NaCl) = \Lambda(NaClO_4) + \Lambda(KCl) - \Lambda(KClO_4) = (117.48 + 149.86 - 140.04)\ \Omega^{-1}\ cm^2\ mol^{-1} = 127.30\ \Omega^{-1}\ cm^2\ mol^{-1}$. This value is in reasonable agreement with the experimental value of 126.45 $\Omega^{-1}\ cm^2\ mol^{-1}$. Of course, this calculation is only possible using the $\Lambda_0$values.

2.3.4 Strong and weak electrolytes

Are there different classes of electrolytes?

Earlier we distinguished electrolytes with different composition and with different charge of the constituting ions. Obviously, this classification is not satisfying if we ask the question:

Why the application of the Kohlrausch's square root law fails in case of acetic acid (Fig. 2.9)?

The answer is given in the following where we distinguish "strong electrolytes" and "weak electrolytes" for which acetic acid is a prominent example.

Acetic acid and many other organic acids, as well as the corresponding salts, are not fully dissociated in aqueous solutions. The concentrations $c(H^+)$ of cations and $c(A^-)$ of anions are lower than expected. The "weak" acid HA dissociates in water not completely. This is described by the following dissociation equilibrium:

$$HA + H_2O \rightleftharpoons A^- + H_3O^+ \tag{2.69}$$

The corresponding equilibrium constant K is given by

$$K = c(A^-)c(H_3O^+)/c(HA)c(H_2O) \tag{2.70}$$

The concentration of water is in very good approximation constant and we get the classical equilibrium constant K_c:

$$K_c = c(A^-)c(H_3O^+)/c(HA) \tag{2.71}$$

It would be more accurate using activities a instead of concentrations, but using concentrations can be accepted because the concentrations of ions are very low and thus the activity coefficients f are close to 1.

In the literature one finds another equivalent formulation of the dissociation equilibrium using the degree α of dissociation. One defines the share of the dissolved, dissociating molecules HA. Starting with a concentration c of HA leads to the concentration $c(A^-) = \alpha \cdot c$ and $c(H_3O^+) = \alpha \cdot c$, where α is the degree of dissociation. The remaining concentration of HA is then $(1 - \alpha)c$. The equilibrium constant K_c is given by

$$K_c = \alpha^2 c^2/(1-\alpha)\, c = \alpha^2 c/(1-\alpha) \tag{2.72}$$

The undissociated molecules HA contribute not to the conductivity. Since in a weak electrolyte according to the dissociation equilibrium only the ions A^- and H_3O^+ contribute to the conductivity the quotient Λ_c/Λ_0 is equivalent to the degree of dissociation α:

$$\Lambda_c/\Lambda_0 = \alpha^- \tag{2.73}$$

Substituting eq. (2.4.23) into eq. (2.4.24) we get Ostwald's "law of dilution":

$$\Lambda_c^2 c/(\Lambda_0 - \Lambda_c)\,\Lambda_0 = K_c \tag{2.74}$$

Here we finish the exploration of macroscopic properties of electrolytes; however, there remain some interesting questions. So far the current came about by the charge transport of both the cations and the anions. Now we are asking: How much one of the ions contributes to the current? How can we find the contribution of single ions to the molar conductivity of the salt?

2.3.5 Transport number and mobility of single ions

To find an answer to this question, we remind ourselves to the beginning of the chapter "Electrolytes" (chapter 2.3). There the velocity of the ions moving in an electric field has been considered. A qualitative answer to our question is: Fast ions transport per unit of time more electrical charge than slow ions, provided the ions have the same charge number. Therefore, if we were able to measure the velocity of the migrating ions, a quantitative answer to our question seems to be possible.

The velocity of colored ions like the violet permanganate ion MnO_4^- can be measured directly. Into a solution containing KNO_3 one carefully layers a solution of potassium permanganate. After applying an electric field one observes the migration of the colored boundary. In an electric field of 1 V/cm the velocity of migration of the MnO_4^- ion is $5 \cdot 10^{-4}$ cm/s.

However, this direct observation is not generally usable, because most of the ions are colorless.

It was Hittorf who propagated a procedure that can generally be used to determine the transport number of ions.

The fraction I^+/I of the current, which is caused by migration of cations, is the transport number t^+ and I^-/I, the fraction of the current I transported by the anions is the transport number t^-.

The transport numbers can be expressed by a number of relations:

$$t^+ = I^+/I = q^+/(q^+ + q^-) = u^+/(u^+ + u^-) = v^+\Lambda^+/(v^+\Lambda^+ + v^{--}\Lambda^-) = v^+\Lambda^+/\Lambda \tag{2.75a}$$

$$t^- = I^-/I = q^-/(q^+ + q^-) = u^-/(u^+ + u^-) = v^- \Lambda^-/(v^+\Lambda^+ + v^{--}\Lambda^-) = v^-\Lambda^-/\Lambda \tag{2.75b}$$

Since q^+ and q^- are the amounts of charge transported by the cations and the anions, the sum of the transport numbers is

$$t^+ + t^- = 1 \tag{2.76}$$

An additional relation is

$$I = u^+/(u^+ + u^-)\, I + u^-/(u^+ + u^-)\, I = v^+\Lambda^+/\Lambda I + v^-\Lambda^-/\Lambda I \tag{2.77}$$

How can we measure t^+ and t^-?

We perform an electrolysis of aqueous hydrochloric acid.

Figure 2.10 shows schematically the experimental setup for measuring the transport numbers. The space of the electrolytic solution is divided into three compartments: The volume near the cathode, the volume near the anode and the "middle volume" between these volumes.

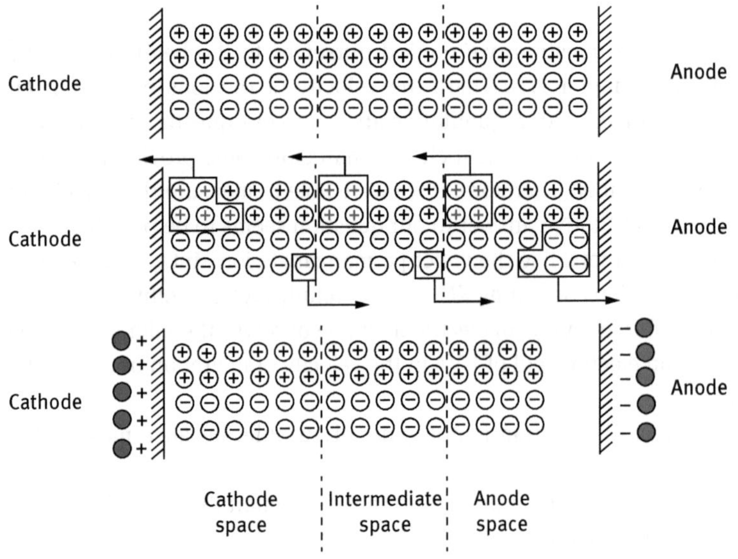

Fig. 2.10: The influence of the different ion mobility on the changes of concentration during an electrolysis according to Hittorf.

We use aqueous HCl and assume that the mobility of the cations H_3O^+ is four times larger than that of the anions Cl^- – what will be corroborated by the experiment. According to eq. (2.77), the cations bring about 4/5 of the current I and only 1/5 of the current is caused by migration of the anions. The current is defined by the number of charges crossing per second the interfaces in the direction of the current. These are the interface between the cathodic volume and the "middle volume" and the interface between the anodic volume and the "middle volume." Simultaneously charge has to be exchanged at the electrodes. For instance, at the cathode by reduction of 5 mol cations and at the anode by oxidation 5 mol of anions. From the "middle volume" migrate 4 mol cations into the cathode volume and 4 mol of cations migrate from the volume near the anode into the "middle volume." Simultaneously 1 mol of anions passes from the "middle volume" into the anode volume and 1 mol of anions migrates from the cathode volume into the "middle volume." This is illustrated in the middle part of Fig. 2.10. The result of the electrolysis is shown in the bottom part of Fig. 2.10. The solution lost 5 mol of HCl, 4 mol is lost in the anode volume and only 1 mol in the cathode. Due to the different mobilities of cations and anions one observes different decreases of the concentrations close to the anode and the cathode after an electrolysis.

The change in the concentration is used for the determination of the transport numbers.

Table 2.3 summarizes the above-described considerations in a general form for the turnover of 1 F in a 1,1 electrolyte. It follows directly:

Table 2.3: Summary of the changes of concentration during the electrolysis of an 1,1 electrolyte and the turnover of 1 F*.

Process	Cathodic space	Middle space	Anodic space
Reaction at the electrode	−1 mol cations		−1 mol anions
Immigration	$+t^+$ mol cations	$+t^+$ mol cations $+t^-$ mol anions	$+t^-$ mol anions
Emigration	$-t^-$ mol anions	$-t^+$ mol cations $-t^-$ mol anions	$-t^+$ mol cations
Change of concentration after turnover of 1 F	$-(1-t^+)$ mol cations $-t^-$ mol anions = = $-t^-$ mol cations − $-t^-$ mol anions = = $-t^-$ mol electrolyte		$-(1-t^-)$ mol cations $-t^+$ molanions = = $-t^+$ mol cations − $-t^+$ mol anions = = $-t^+$ mol electrolyte

* Taken from G. Wedler, H.-J. Freund, "Lehrbuch der Physikalischen Chemie", 6th ed. Wiley-VCH 2012.

$$\Delta c_{\text{anode}}/\Delta c_{\text{cathode}} = u^-{}_{\text{anion}}/u^+{}_{\text{cation}}. \tag{2.78}$$

and

$$\Delta c_{\text{anode}}/\Delta c_{\text{cathode}} = t^+/t^- \tag{2.79}$$

and with $t^+ + t^- = 1$

$$t^{+} = \Delta c_{\text{anode}} / \Delta c_{\text{anode}} + \Delta c_{\text{cathode}} \text{ and } t^{-} = \Delta c_{\text{cathode}} / \Delta c_{\text{anode}} + \Delta c_{\text{cathode}} \quad (2.80)$$

The transport numbers can be calculated by the change in concentration, provided the change in the concentrations near the anode and the cathode is not falsified by diffusion or stirring.

For clearness a short remark is useful. In the electrolysis, the time, that is, the intensity of the current, is not relevant. We are only interested in the transferred charge independent from the time which is used for it. In a later section, we will again discuss transport problems, but then in connection with the intensity of the observed current and we will find that contributions of diffusion to the mass transport cannot be excluded a priori.

Measuring at concentrations lower than 0.01 mol delivers in good approximation the transport numbers $t^{+} = t_0^{+}$ and $t^{-} = t_0^{-}$ at "infinite dilution" from which with Λ_0 one obtains the limiting conductivities Λ_0^{+} and Λ_0^{-}.

Transport numbers can be measured by several approaches and numerous data for solutions appear in the literature. The few data displayed in Tables 2.4 and 2.5 result from the application of the relations and experimental methods described in the preceding text.

Table 2.4: Hittorf transport numbers in aqueous solution at 298 K.

	HCl	LiCl	NaCl	KCl	KBr	KJ	KOH	$CaCl_2$	$LaCl_3$	K_2SO_4
t^{+}	0.821	0.337	0.401	0.496	0.484	0.489	0.274	0.438	0.477	0.477
t^{-}	0.179	0.663	0.599	0.504	0.516	0.511	0.726	0.562	0.523	0.523

Table 2.5: Molar limiting conductivity Λ_0^{+} and Λ_0^{-} of single ions in aqueous solutions at 298 K.

Cation	H^{+}	Li^{+}	Na^{+}	K^{+}	Rb^{+}	Cs^{+}	Ag^{+}	NH_4^{+}
Λ_0^{+}	349.8	38.7	50.1	75.4	77.8	77.2	61.9	73.6
Cation	$N(CH_3)_4^{+}$	$N(C_2H_5)_4^{+}$	$N(C_3H_7)_4^{+}$	$N(C_4H_9)_4^{+}$				
Λ_0^{+}	44.9	32.7	23.4	19.5				
Cation	½ Mg^{2+}	½ Ca^{2+}	½ Sr^{2+}	½ Ba^{2+}	½ Cu^{2+}			
Λ_0^{+}	53.1	59.1	59.5	63.6	56.6			
Anion	OH^{-}	F^{-}	Cl^{-}	Br^{-}	J^{-}	NO_3^{-}	BrO_3^{-}	ClO_4^{-}
Λ_0^{-}	198.6	55.4	76.4	78.1	76.8	71.5	64.6	55.7
Anion	½ SO_4^{2-}	½ CO_3^{2-}	$HCOO^{-}$	CH_3COO^{-}	$C_2H_5COO^{-}$			
Λ_0^{-}	80.0	69.3	54.6	40.9	35.8			

The ions H^+ and OH^- show outstanding values. The values of the tetraalkyl ammonium ions reveal clearly the influence of the size on the ion mobility; the larger the ion, the lower the mobility in contrast to the alkali ions where the lithium ion has a remarkably low value. This is explained by the large solvation shell of this ion.

2.3.6 Dependence of the conductivity from temperature and solvent

So far we described the conductivity of electrolytes at room temperature.
Considering the relation

$$\Lambda_{0i}{}^{\pm} = F\, z_i u_{0i}{}^{\pm} = F \cdot z_i^2 \cdot e/\ 6\pi r_i \cdot \eta \tag{2.81}$$

the dependence of the conductivity from temperature and solvent can be traced back to the viscosity, provided the radius of the solvated ion remains unchanged. This means that the time-averaged solvation shell in the range of the temperature interval under consideration is neither increased nor decreased and that the radius of the ion depends not on the solvent. Tantamount to that the product $\Lambda_{0i}{}^{\pm} \cdot \eta$ should not depend on the temperature.

$$\Lambda_{0i}{}^{\pm} \cdot \eta = F \cdot z_i^2 \cdot e/6\pi\, r_i \tag{2.82}$$

Equation (2.82) is known as Walden's rule.

The data of the following table (2.6) shows that Walden's rule is approximatively fulfilled for ions with strong binding of the solvent molecules (Li^+) and for large ions ($N(C_2H_5)_4{}^+$) with weak interaction with the solvent molecules.

During the last years more and more electrochemical experiments have been performed in nonaqueous solvents.

Table 2.7 shows the influence of different solvents on the molar conductivity of K^+, $N(C_5H_{11})^+$ and the picrate anion and on the validity of Walden's rule.

In the following table 2.8, the radii of atoms and ions are compiled as obtained from X-ray investigations. In addition, the radii of some solvated ions and their mobility are shown. The data document the big differences due to the solvation of ions.

The large dipole moment of the water molecule leads to the solvation of the ions by electrostatic ion–dipole interaction. The radii of the solvated ions, calculated from their mobility, are larger than the ionic radii or the atomic radii; this arises because solvent molecules form a solvation shell around the ion. The solvation shell increases with increasing field strength at the surface of the central ion. Therefore, in aqueous solution solvated Li^+ and F^- have the largest "solution radii" and the smallest mobility. The ratio of the ionic radii of nonsolvated Li^+ and K^+ is about 1:2, which means the electric field strength at the boundary of the Li^+ ion is four

In the past, considerable efforts have been made to establish models of electrolyte solutions which are in agreement with the properties obtained from experiments. An early standard model is that presented by Debye and Hückel, which will shortly be described in the following.

2.3.7 Ionic interaction in electrolyte solutions

The decreasing molar conductivity with increasing concentration of the electrolyte points to a decreasing density of ionic charges in the solution. As it is observed with "strong electrolytes," which are completely dissociated in ions at low concentrations, it is reasonable to assume that increasing ion–ion interaction becomes more important at larger concentrations. Cations and anions may form ion pairs and larger aggregates.

The energy of interaction of two neutral molecules falls off as R^{-6}, where R is their distance, but the Coulomb interaction between two charged ions falls of by R^{-1}. Ionic solutions deviate sharply from ideality on account of the long distance of their electrostatic interactions. This indicates not only that ions interact over long distances, but also that the deviations from ideality are likely to be dominated by the direct electrostatic interactions between solvated ions. It is well known that activities a, which are used in case of ionic solutions instead of concentrations c, take into account this interaction. Concentration and activity are connected by the relation

$$c_i = f_i\, a_i \tag{2.83}$$

where f_i is the activity coefficient.

In analogy to eq. (2.83) can be formulated:

$$\Lambda_c = f_\Lambda\, \Lambda_0 \tag{2.84}$$

where f_Λ is the conductivity coefficient.

Is the activity coefficient just an empirical adaption of activity and concentration or is there a story behind the story? It seems there is a story and we will enlighten it. The title of the story is "Debye–Hückel model"; it will be the subject of the following section.

2.3.8 The Debye–Hückel model

Peter Debye and Erich Hückel were looking for a model of the ionic structure in solution. In further steps an attempt has been made to find a correlation between the conductivity coefficient f_Λ and the interionic force following a model for the electrostatic interaction in ionic solutions.

Peter Debye and Erich Hückel published their model in 1923. It is an excellent example how physical features are discerned and then incorporated into a quantitative model. Every step is guided by the underlying physical ideas.

The calculation of the activity coefficient is one of the topics of the Debye–Hückel theory and in the center of interest.

Debye and Hückel made the following approximations simplifying the problem:
- For the interactions of the ions, Coulomb forces alone are responsible; all the other intermolecular forces are neglected.
- The dielectric constant ε of the solution is replaced by that of the pure solvent, that is, changes of ε by the dissolved salts are not considered.
- The ions are spherical, nonpolarizable charges with a spherically symmetric electric field.
- The energy of attraction due to the interaction of the ions is taken to be small in comparison to the thermal energy.
- Strong electrolytes are considered to be completely dissociated at all concentrations.

Opposite charged ions attract each other. This suggests that cations and anions are not uniformly distributed, but that more anions tend to be found in the vicinity of cations and vice versa (Fig. 2.11). Overall the solution is neutral, but in the vicinity of any given ion there is a predominance of ions of opposite charge by the counter-ions. On average more counter-ions than like-ions pass by any given ion, and they come and go in all directions. This time-averaged spherical haze of opposite charge is called the ion cloud (sometimes called the ionic atmosphere).

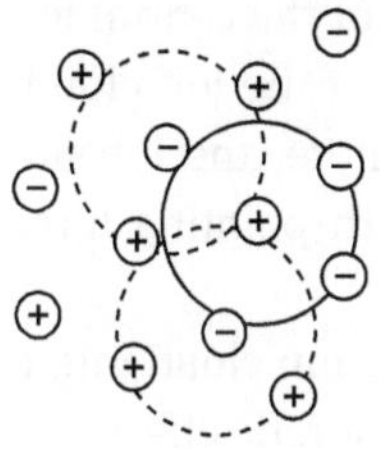

Fig. 2.11: Near-order in solutions of electrolytes (ionic cloud).

The radii of the ion clouds increase strongly with decreasing concentration and decreasing charge of the ions. In 10^{-1} M solutions the radii of the ion cloud are of the order of the ion radii, in 10^{-4} M solutions they are about two orders larger.

The values of R in Table 2.9 have been calculated with the following equation:

$$R(\mathrm{H_2O},\ 298\mathrm{K}) = 1.358 \cdot 10^{-8}\,(\mathrm{mol/m})^{1/2}\left(\sum z_i^2 c_i\right)^{1/2} \tag{2.85}$$

Table 2.9: Radius R of the ion cloud for aqueous solutions of different types of electrolytes at 298 K.

c (mol/L)	*R*(1,1) (nm)	*R*(1,2) (nm)	*R*(2,2) (nm)	*R*(1,3) (nm)
10^{-1}	0.96	0.55	0.48	0.39
10^{-2}	3.04	1.76	1.52	1.24
10^{-3}	9.6	5.55	4.81	3.93
10^{-4}	30.4	17.6	15.2	12.4

Due to the attracting and repelling electrostatic forces, a short-range order is developed. Every ion is simultaneously central ion and component of the ion cloud around a neighboring ion. The solvent molecules and ions are subject to the thermal motion, which counteracts the formation of the near-order.

The energy and therefore the chemical potential of the central ion is lowered by its favorable electrostatic interaction with its ionic atmosphere. The main task is to find a way of formulating this effect quantitatively.

The Coulomb attraction is taken to be small in comparison to the thermal energy of motion. This assumption is justified for diluted solutions. The average distance of ions in such solutions is large, so that the electrostatic interaction which decreases with the square of the distance is reasonably low. In this diluted solutions, the permittivity ε of the solutions is practically identical with that of the solvent.

If an outer electric field is applied to the solution the ions start migration. Due to their different charge, the ions move in opposite directions to the electrodes. This leads to two additional effects: the *relaxation effect* and the *electrophoretic effect*.

The origin of the relaxation effect is the opposite acceleration of the central ion and the ions in the ion cloud. This leads to a partial destruction of the ion cloud, and the rearrangement of the ion cloud needs some time. Therefore, the ion migrates ahead of the center of the ion cloud. This leads to a return pointing force slowing down the movement of the ions.

The opposite movement of the central ion and the surrounding ion cloud intensifies the effect of friction in the solvent. This is called the electrophoretic effect.

The basic equations used in the Debye–Hückel theory are the Poisson equation (see Box 4) and the “Boltzmann distribution.”

Box 4
Application of the Poisson equation

The Poisson equation describes the relation between the electrical potential φ and the charge density ρ. It is given by

$$\Delta\varphi = -4\pi\rho \tag{1}$$

where Δ is the Laplace operator $\Delta = \partial^2/\partial x^2 + \partial^2/\partial y^2 + \partial^2/\partial z^2$ in Cartesian coordinates.

The parallel plate condensers is an example of how to use the Poisson equation.

In a figure, P_1 and P_2 are the large plates of the condenser, and d is the distance between the plates. The metallic plates are equipotential surfaces. The electric field strength E is directed perpendicular to the surface of the plates. At the surface of the plates, the charge density η exists. The space between the plates is free of charge; therefore, it is

$$\text{div}E = \Delta\varphi = 0 \tag{2}$$

the potential depends only on the x coordinate. The potential equation for this case is:

$$d^2\varphi/dx^2 = 0 \tag{3}$$

and the integration results in the linear relation

$$\varphi = ax + b \tag{4}$$

to determine a and b, we use the surface charge density η.

The power flux crossing an area ΔF is directed exclusively into the interior of the condenser, thus we get

$$\varphi = x(df) = 4\pi\eta\, df \tag{5}$$

$$d\varphi/dx = -4\pi\eta \tag{6}$$

and as eqs. (4) and (6) follows from P_1

$$d\varphi/dx = a = -4\pi\eta \tag{7}$$

The constant b remains undefined, because we can put on P_2 an arbitrary potential, for example, the potential zero by earth connection. The potential equation is

$$\varphi = -4\pi\eta x + b$$

From this equation we get the following relations:

$$x = 0; \quad \varphi_1 = b$$

$$x = d; \quad \varphi_2 = -4\pi\eta d + \varphi$$

$$-x = x; \quad \varphi = \varphi_1 - 4\pi\eta x$$

$$\varphi_1 - \varphi_2 = U = 4\pi\eta d$$

The field strength E is in the space between the plates constant (homogeneous field) and has the value

$$E_x = U/d$$

The capacity C of the condenser can be calculated with

$$C = e/U = \eta F/4\pi\eta d = F/4\pi d$$

if the dimensions of the condensor are known.

The Poisson equation – here formulated in spherical coordinates:

$$1/r^2 \cdot \delta\{r^2\, \delta\psi(r)/\delta r\}/\delta r = -\rho(r)/\varepsilon\varepsilon_0 \tag{2.86}$$

describes the relation between the electrical potential $\varphi(r)$ and the density of charge $\rho(r)$ in the distance r from the central ion. ε is the permittivity of the solvent and ε_0 is the electrical field constant.

As mentioned earlier, in the solution of electrolytes exists the competition between the electrostatic interaction and the thermal energy.

The key for reaching a first goal is the ratio between the number ${}^1N_i(r)$ of ions per volume unity in the distance r from the central ion and ${}^1\tilde{N}_i$ the averaged number density of these ions in the volume of the solution. According to the "Boltzmann distribution," this is

$${}^1N_i(r)/{}^1\tilde{N}_i = e^{-z_i \cdot e \cdot \psi(r)/kT} \tag{2.87}$$

In the following, we omit the elaborated pathway to the central equation (2.88), it is described in the Appendix E.

Here we continue with the discussion of the resulting central equation, which allows a quantitative description of the interionic interaction in the frame of the Debye–Hückel model.

The central equation of the Debye–Hückel theory is

$$\psi(r) = (z_i\, e/4\pi\,\varepsilon\varepsilon_0)\ (e^{a/\beta}/1 + a/\beta)\ (e^{-r/\beta}/r) \tag{2.89}$$

where $1/\beta = (2\, N_A\, e^2\, I/\varepsilon\varepsilon_0\, kT)^{1/2}$, a is the radius of the central ion, $I = \Sigma z_i^2 c_i$ is the ion strength. The discussion of this equation is of some interest. We distinguish two contributions to the potential $\cdot\varphi(r)$, one $\cdot\varphi_z(r)$, due to the contribution of the central ion and a second one, $\cdot\varphi_w(r)$, due to the contribution of the ion cloud.

The first one is given by

$$\psi_z(r) = z_i\, e/4\pi\,\varepsilon\varepsilon_0\, r \tag{2.90}$$

In the case of nonexisting interionic interaction this equation would still be valid. The genuine ionic interaction is given as follows:

$$\psi_w(r) = \psi(r) - \psi_z(r) = (z_i\, e/4\pi\,\varepsilon\varepsilon_0\, r)\ (e^{a/\beta}/1 + a/\beta)\ (e^{-r/\beta} - 1) \tag{2.91}$$

where β is the dominating parameter. β is inversely proportional to the square root of the ion strength I. If the concentration just as the ion strength strive to zero, β moves versus infinity. Then, according to eq. (2.91), $\varphi_w(r)$ moves versus zero.

The radius of the ion cloud

β has the dimension of a length and is called the radius of the ion cloud, also called the "Debye length" or "shielding length."

At this distance from the center of the central ion the charge density $d\rho/dr = \rho\, 4\pi\, r^2$ is at maximum. This becomes evident from the following equation:

$$\rho(r) = -\,(z_i\, e/4\pi\, \beta^2)(e^{a/\beta}/1 + a/\beta)(e^{-r/\beta}/r) \tag{2.92}$$

By multiplication with $4\pi\, r^2 dr$ we calculate the charge in a spherical shell with the radius r and the thickness dr:

$$\rho 4\pi\, r^2\, dr = \text{const.}\ r\, e^{-r/\beta}\, dr \tag{2.93}$$

The extreme value of this function is obtained by differentiation and equating the first derivative of this function to zero:

$$d\,(\rho 4\pi\, r^2)/dr = 0 = \text{const.}(e^{-r/\beta} - r/\beta\ e^{-r/\beta}) \tag{2.94}$$

This equation fulfils

$$\beta = r(\rho 4\pi\, r^2)_{\text{max}} \tag{2.95}$$

Putting the universal constants into eq. (2.95) yields for the radius β of the ion cloud:

$$\beta = 6.288 \cdot 10^{-11} (\text{mol/Km})^{1/2}\, (\varepsilon_T T/I)^{1/2} \tag{2.96}$$

$$\beta(\text{H}_2\text{O},\ 298\ \text{K}) = 1.358 \cdot 10^{-9}\, (\text{mol/m})^{1/2} \left(\sum z_i^2 c_i\right)^{-1/2} \tag{2.97}$$

With eq. (2.97) the values of the radius of the ion cloud given in Table 2.9 have been calculated.

Calculation of the conductivity coefficient f_Λ

This calculation is an extension of the Debye–Hückel model by Onsager. It is rather complicated and therefore only the principles shall be outlined briefly.

The molar conductivity Λ_c consists of three contributions: the molar limiting conductivity Λ_0, the contribution of the relaxation effect Λ_{rel} and the contribution of the electrophoretic effect Λ_{el}.

$$\Lambda_c = \Lambda_0 - \Lambda_{\text{rel}} - \Lambda_{\text{el}} \tag{2.98}$$

For an 1,1 electrolyte ($z^+ = 1$, $z^- = -1$) one obtains the following equation:

$$\Lambda_c(1,1) = \Lambda_= - (7.281 \cdot 10^{-3}\, \Lambda_0 + 1.918 \cdot 10^{-4}\, \text{m}^2\, \Omega^{-1} \text{mol}^{-1})(\text{m}^3/\text{mol})^{1/2}\, c^{1/2} \tag{2.99}$$

and for the conductivity coefficient f_Λ

$$f_\Lambda(1,1) = 1 - -1/\Lambda_0(0.2302\, \Lambda_0 + 60.68\ \text{cm}^2\, \Omega^{-1}\, \text{mol}^{-1})(\text{dm}^3/\text{mol})^{1/2} c^{1/2} \tag{2.100}$$

The agreement of eq. (2.100) with the square root law of Kohlrausch is obvious.

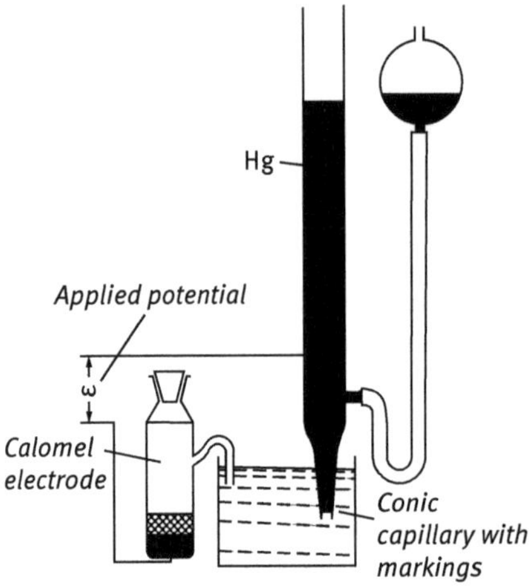

Fig. 2.13: Lipmann's experimental setup for measuring the electrocapillarity of mercury.

changes the surface tension of the mercury. At every voltage the meniscus of the mercury in the capillary is brought in line with the mark by changing the heights of the mercury in the pipe. The height of the mercury is proportional to the surface tension of mercury. In this way, the dependence of the surface tension σ of mercury on the applied voltage has been registered. The electrocapillarity curves are parabolas as shown in Fig. 2.14 The voltage at the maximum of the curve is called "electrocapillary maximum" and is the "potential of zero charge" (PZC).

First, a qualitative explanation of the electrocapillarity will be given. At the uncharged surface of the metal, the surface forces make the surface as small as possible.

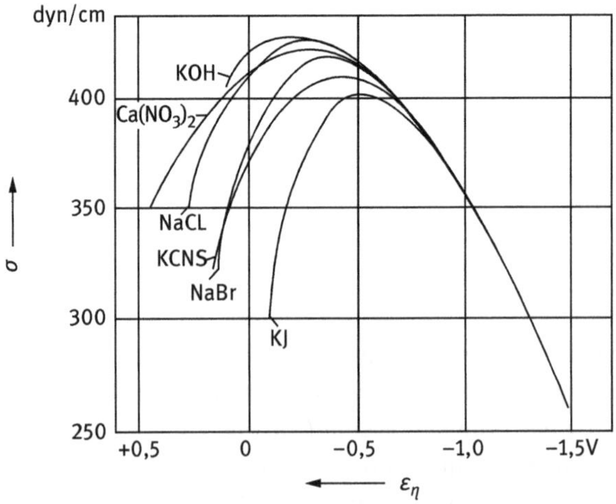

Fig. 2.14: The surface tension σ of mercury contacting aqueous solutions of different salts depending on the voltage ε (ε is measured vs the NHE at 18 °C).

Positive as well as negative electric charges at the surface repel each other. They tend to extend on a surface as large as possible, thereby enlarging the surface. Thus, charging the surface counteracts the surface tension. This effect is independent from the sign of the electrical charge and it is the larger the electric charging is.

The surface is obviously a capacitor. The surface tension σ is reduced by the energy E/cm^2 which is used for loading of the double layer capacitor with the capacity C_d.

The relations between the energy E, the surface tension σ and the capacity C_d describes the following equation:

$$dE = -d\sigma = \varepsilon C_d \, d\varepsilon = Q \, d\varepsilon \tag{2.107}$$

where ε is the voltage applied to the mercury surface and Q (Cb/cm^2) is the charge density at the metal surface.

The Lippman–Helmholtz equation is

$$d\sigma/d\varepsilon = -Q \tag{2.108}$$

and further differentiation yields:

$$d^2\sigma/d^2\varepsilon = -dQ/d\varepsilon = C_d \tag{2.109}$$

The experimental electrocapillarity deviates from the ideal form of a parabola and in addition the maximum of the curves, where no excess charge exists at the surface, is shifted in the different electrolytes. This has been explained by the adsorption of ions or neutral molecules.[9]

The investigations on the mercury/electrolyte system revealed the existence of an electrified interface and its basic properties. For many decades, mercury was the preferred metal for electrochemist. Its surface is perfectly flat and is easily renewable.

Of course electrochemistry is not limited to mercury but many other solid electrodes are used. For studying the electric double layer of solid electrodes nowadays mostly the capacity of the electrode is measured.

Considering the importance of this electrified interphase, we explain the double layer in more details.

2.4.2 Theory and models of the double layer

Helmholtz was the first who concluded from Lipmann's experimental results that the formation of a potential difference at the interphase rises from positive and negative excess charges at the phase boundary. This simple idea on the distribution of ions in the double layer electrode/electrolyte has been published by

9 M. Gouy, Ann. Chi. Phys. 8, 291 (1906) and 9, 75(1906); N.W. Nikolajewa, A.N. Frumkin, S.A. Jopa, J. Phys. Chem USSR, 26, 1326 (1952).

Helmholtz.[10] He invented the picture of a parallel plate condenser for the interface (Fig. 2.15) of a metal ion electrode.

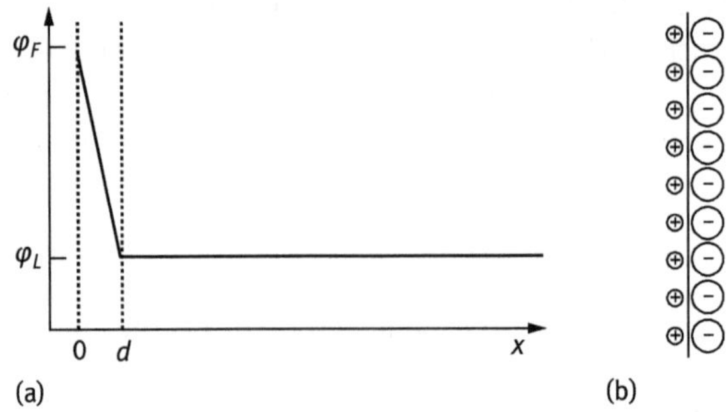

Fig. 2.15: The dependence of the potential from the distance x and the charge distribution in the double layer according to the model of Helmholtz.

As shown in Fig. 2.12 at a metal/ion interface, metal ions are exchanged across the phase boundary between the electrode and the electrolyte. If cations leave the metal, a negative charge remains at the metal surface. Helmholtz assumed that the negative excess charge at the metal surface brings about a layer of cations at the electrolyte site of the phase boundary compensating the negative excess charge of the metal surface. This double layer can indeed be described by the equivalent circuit of a parallel plate condenser with the integral capacity K:

$$K = q^{M} / \Delta E \tag{2.110}$$

where q^M is the excess charge at the metal side and ΔE is the potential difference across the double layer. The differential capacity C in this model comes to

$$C = \varepsilon / 4\pi d \tag{2.111}$$

ε is the permittivity of the medium (generally used that of the solvent) and d is the distance between the plates of the "condenser." It is the distance between the centers of the opposite charges.

The Helmholtz model meets not the experimental results measured at the electrolytic double layer.

One has to take into account the thermal energy of the ions. This leads from the rigid layer of Helmholtz to a diffuse layer, where the rigid layer is permanently destroyed and recovered. This model has been proposed Gouy[11] and Chapman[12].

10 H. Helmholtz, Wied. Ann. J., 337 (1879).
11 A. Gouy, J. Phys. 9, 457 (1910), Compt. reud. 149, 654 (1909).
12 D. L. Chapman, Phil. Mag. 25, 475 (1913).

For the space charge density ρ and the potential φ exist two defining equations. The Boltzmann equation:

$$\rho = F \sum z_i c_i \exp(z_i F \Delta\varphi / RT) \tag{2.112}$$

where $\Delta\varphi$ is the potential difference between the surface of the electrode and the interior of the electrolyte, z_i is the charge number of the ions taking into account their sign, c_i is the concentration of the ions i in the bulk of the electrolyte.

Simultaneously, the Poisson equation is valid:

$$\partial^2\varphi/\partial x^2 + \partial^2\varphi/\partial y^2 + \partial^2\varphi/\partial z^2 = -4\pi/D\rho \tag{2.113}$$

where D is the dielectricity constant, and x, y, z are space coordinates.

Combination of eq. (2.112) with eq. (2.113) yields the differential equation for $\Delta\varphi$:

$$\partial^2\varphi/\partial x^2 + \partial^2\varphi/\partial y^2 + \partial^2\varphi/\partial z^2 = -4\pi F/D \cdot \sum z_i c_i \exp(z_i F \Delta\varphi / R T) \tag{2.114}$$

Equation (2.114) is identical with the equation used later by Debye and Hückel for calculating the charge distribution in the ion cloud and the activity coefficient of ions in the solution (Fig. 2.16).

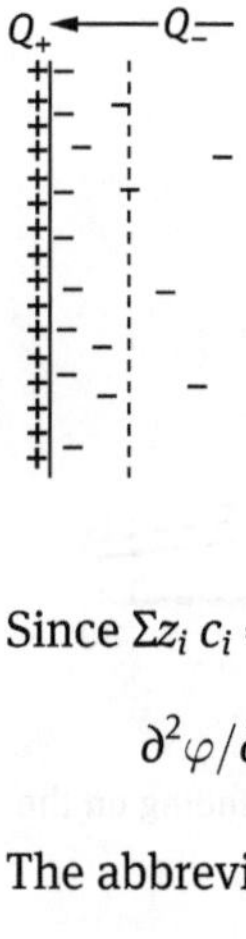

Fig. 2.16: Schematic illustration of the charge distribution in the diffuse layer.

Since $\Sigma z_i c_i = 0$ (electroneutrality) for $z_i F\Delta\varphi/RT \ll 1$, eq. (2.114) transforms into

$$\partial^2\varphi/\partial x^2 + \partial^2\varphi/\partial y^2 + \partial^2\varphi/\partial z^2 = 4\pi F^2/DRT \sum z_i^2 c_i \cdot \Delta\varphi = \kappa^2 \Delta\varphi \tag{2.115}$$

The abbreviation κ defines the reciprocal distance from the electrode surface:

$$\kappa = (4\pi F^2/DRT \sum z_i^2 c_i)^{1/2} \tag{2.116}$$

For a planar surface we get due to $\partial^2\varphi/\partial y^2 + \partial^2\varphi/\partial z^2 = 0$:

$$\Delta\varphi = \varphi_0 \exp(-\kappa x) \tag{2.117}$$

where φ_0 is the potential at the origin of the diffuse layer, that is, at the surface of the electrode, or better at the plane of nearest approach of the ions (Fig. 2.17).
The width ∂_d of the double layer is approximately represented by $1/\kappa$.

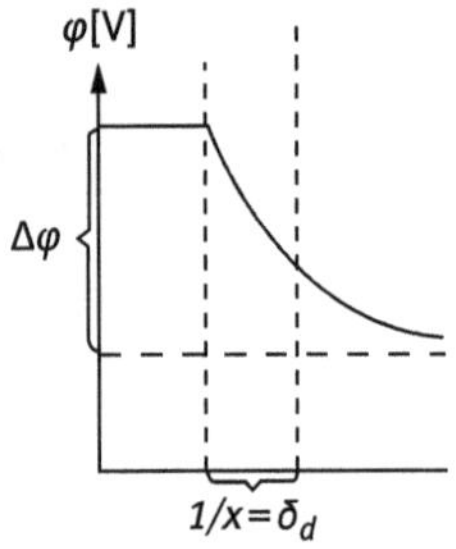

Fig. 2.17: Decay of the potential φ in the diffuse double layer according to the model of Gouy and Chapman.

The exponential decay of the potential is increasing with increasing φ_0, the steeper the decay the smaller is the thickness of the diffuse layer (Fig. 2.18).

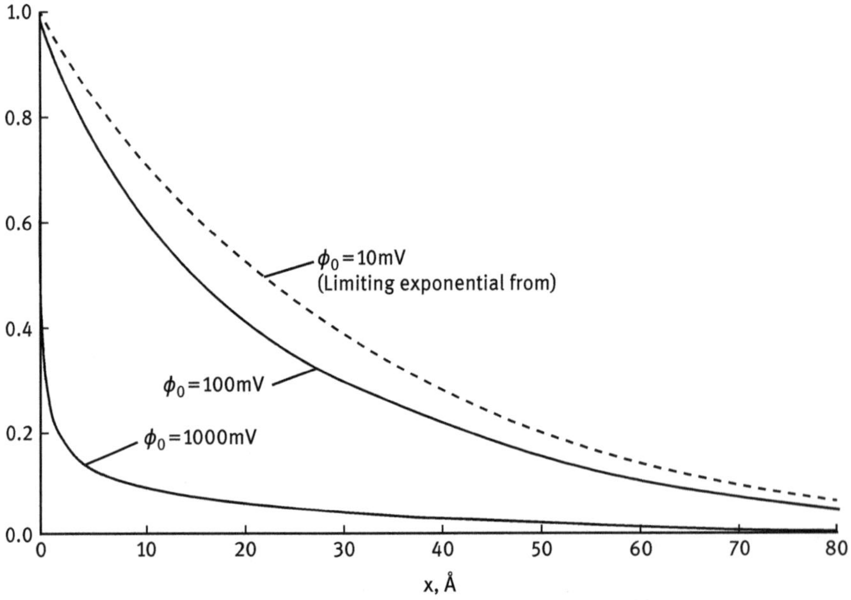

Fig. 2.18: Extension of the diffuse double layer into the interior of the electrolyte depending on the value of the potential φ_0.

2.4.2.1 The excess charge Q_e

Due to the sign of the voltage Φ_0, cations or anions from the interior of the electrolyte are attracted and lead to an excess charge Q_e near the electrode surface:

$$Q_e = -(D\kappa/4\pi)\Phi_0 \tag{2.118}$$

Implementing this expression into eq. (2.117) yields

$$\Delta\varphi = -(4\pi/D\kappa)Q_e\exp(-\kappa\,x) \tag{2.119}$$

In Fig. 2.19, the concentration of anions and cations is shown. Since Φ_0 is positive (+15 mV), the concentration of anions is higher than that of cations, but in the diffuse layer as well anions and cations are enriched.

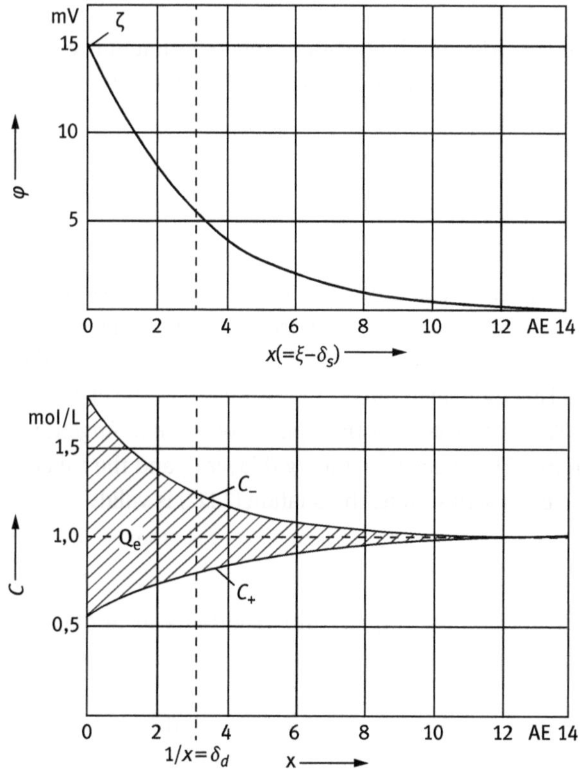

Fig. 2.19: Course of the potential and the concentration in the diffuse double layer for a 1,1 electrolyte (c = 1 mol/L, φ_0 = 15 mV, T = 25 °C).

2.4.2.2 The capacity C_d of the diffuse double layer

The capacity of the diffuse double layer follows from eqs. (2.117) and (2.119). Provided $z_iF\Delta\varphi/RT \ll 1$ is valid C_d comes to

$$C_d = -\,dQ_e/d\varphi_0 = (D/4\pi)\kappa = (DF^2/\,4\pi RT\sum z_i^2c_i)^{1/2} \tag{2.120}$$

The capacity of the diffuse layer depends strongly on the concentration of the electrolyte.

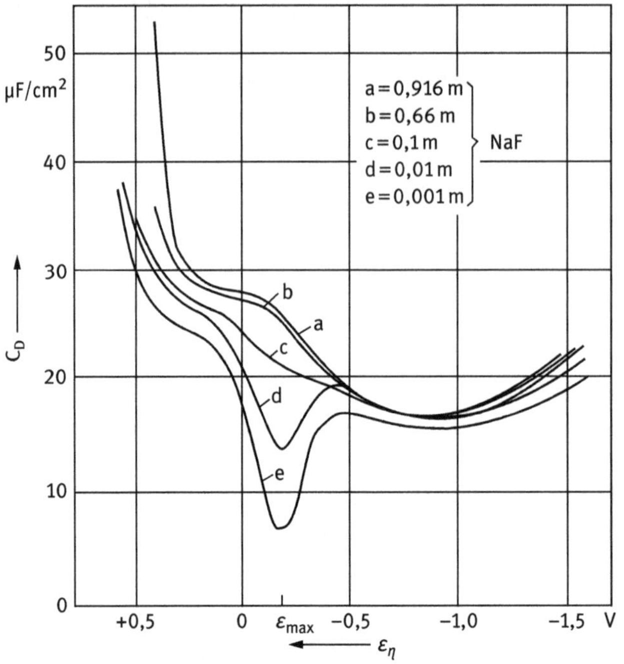

Fig. 2.21: Dependence of the differential capacity C_d of mercury on the potential and at different concentrations of the electrolyte NaF (ε vs NHE, 25 °C).

By charging the electrode, in the phase boundary develops field strengths in the order of 10^5–10^7 V/cm. This leads to the question: How are dipolar water molecules influenced by this strong electrical field?

In liquid water exists a network of water molecules connected by hydrogen bridges forming bunches with tridymite-like structure. The energy of the hydrogen bond between water molecules is much lower than conventional chemical bonds and therefore water molecules are exchanged frequently from their position. This allows the assumption that due to their dipole moment, water molecules in contact with the surface are highly oriented in the electrical field (dielectric saturation); the permittivity in the range of the rigid double layers is much lower (about 5–10) than in bulk water (80) and it depends on the distance from the electrode.

A considerable number of theories and models has been published to explain the experimental results. Reviewing the complete relevant literature is out of the scope of this chapter.

Here we will only mention as a conclusion an often used model developed by Bockris, Devanathan and Miller (BDM model, fig. 2.22). It was one of the milestones in modeling the phase boundary metal/electrolyte.

The BDM model takes into account two different orientations of the water molecules at the surface. Depending on the potential, the dipole of the water molecule can be

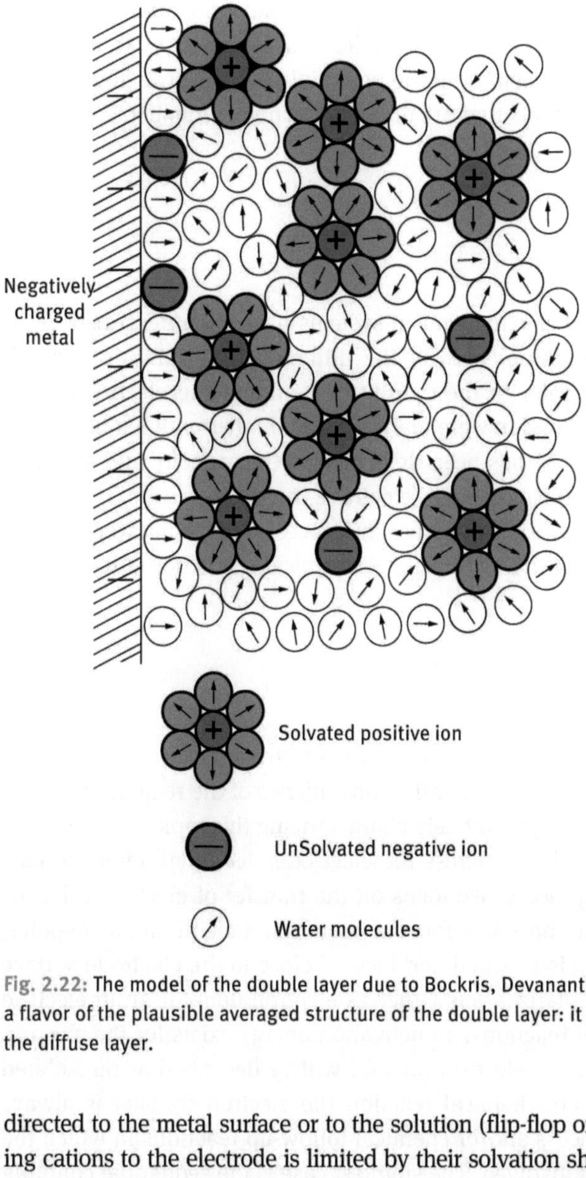

Fig. 2.22: The model of the double layer due to Bockris, Devananthan and Miller (BDM model) gives a flavor of the plausible averaged structure of the double layer: it underestimates the dynamics in the diffuse layer.

directed to the metal surface or to the solution (flip-flop or two-state model). Approaching cations to the electrode is limited by their solvation shell. Specific adsorbing anions (iodide, etc.) are partially in direct contact with the metal surface. The authors suggest different values for the permittivity depending on the distance from the electrode.

Based on this model, the reason for the "hump" in the capacity voltage curves is explained by reorientation of the water dipoles. The small influence of the size of cations is explained by their large distance from the electrode due to solvation. The BDM model agrees qualitatively with the experimental observations; however, it should be noticed that in the double layer exists high mobility of the ions and solvent molecules. The picture shown in Fig 2.22 is an instantaneous snapshot.

Finally, it should be emphasized that the simplest case of the double layer has been described so far. The influence of dipole molecules, for example water, and of adsorption of ions or neutral molecules on the potential of the electric double layer leads to more sophisticated models that are out of the scope of this book. They can be found in the literature and in textbooks.

2.5 Kinetics of electrode processes

In the preceding sections, the electrochemical systems have been described in the equilibrium state (Nernst equation) and by the difference in equilibrium states. Data like standard potential refer to the current-free state of the electrochemical cell. The models and their discussion were mainly based on thermodynamic considerations. In addition, chapter 2.4 has been dedicated to the capacity and structure of the electric double layer of the electrolyte/electrode interface.

Now we make the next step, we will discuss the processes occurring when *current flows* in an electrochemical cell, that is, the electrochemical cell is not in the equilibrium state.

2.5.1 Some preliminary remarks

Describing the kinetics of charge transfer across the electrode/electrolyte interface, we get to the heart of electrochemistry. For the convenience of the reader, some preliminary remarks shall be made qualitatively characterizing this topic.

The transfer of electrical charge across the electrode/electrolyte interface may be described as an isolated process. We focus on the transfer of electrons, that is, on redox electrodes. The electron transfer occurs between a metal and a "depolarizer" (molecule or ion) which is solvated and located close to the electrode surface in the double layer. The depolarizer acts either as electron donator or an electron acceptor. Similar to chemical reactions, an activation energy exists for the electron transfer. In the next chapter, the electron transfer will be described as an isolated process. However, in an electrochemical reaction the electron transfer is always coupled with a transport process and/or chemical follow-up reactions in which the electroactive "depolarizer" is involved. The simplest case is the sequential coupling of electron transfer and mass transport between the electrode surface and the bulk of the electrolyte. The rate of the charge transfer is influenced by the potential at the electrode, whereas the transport rate by diffusion depends only on a concentration gradient. As in conventional chemical kinetics, the slowest step in the line of sequential steps controls the velocity of the complete process. This means that at low currents the charge transfer across the interface is rate determining, whereas at fast charge transfer the slower transport, preferably diffusion, dominates.

Figure 2.23 represents the currents on a redox electrode observed under different conditions. The redox reaction is

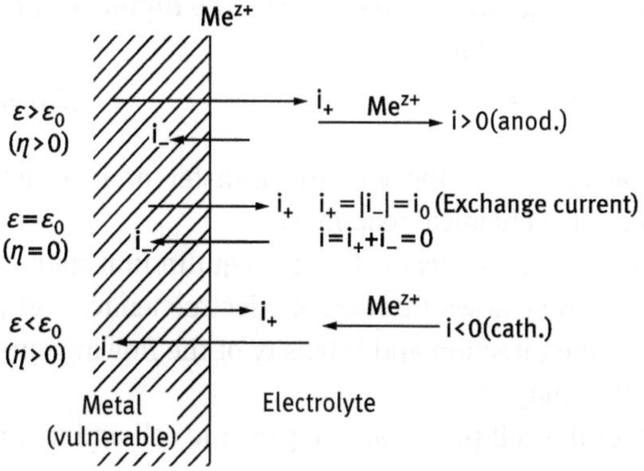

Fig. 2.23: Charge exchange at a redox electrode. Three different situations are shown in the figure: oxidation of S_r ($i_+ > |i_-|$, on top) ; equilibrium at the electrode (middle); reduction of S_o ($|i_-| > i_+$, below).

$$S_r \rightleftharpoons S_o + e^- \tag{2.123}$$

where $S_{r\text{-}}$ are the reduced species and S_o the oxidized species.

The direction of the arrows symbolizes the flux of the electrons and their length the intensity of the current. The observable current I is given by the sum of an anodic current i_+ and a cathodic current i_-:

$$I = i_+ + i_- \tag{2.124}$$

In the center of the scheme the equilibrium situation is depicted. Both the anodic and the cathodic current compensate each other. No current is observed and the overvoltage is zero at the equilibrium state. The currents i_+ and $|i_-|$ at the equilibrium state are called exchange currents.

At a more positive potential dominates the anodic current and at more negative potentials dominates the cathodic current.

The redox reaction occurs in the double layer; electrons are transferred across the boundary metal/electrolyte.

2.5.2 Overvoltage

For the first time, the term overvoltage η appears in Fig. 2.23. Overvoltage is very important in describing electrochemical kinetics; therefore, some remarks are made on it here.

Overvoltage η is the deviation of the electrode potential from the equilibrium potential ε_0 . The exchange current i_0 is the current flow in an electrochemical cell at the equilibrium potential ε_0.

Thus, the overpotential and overvoltage, respectively, show the difference between the actual potential ε and the equilibrium potential ε_0:

$$\eta = \varepsilon - \varepsilon_0 \tag{2.125}$$

Overvoltage may be caused by the barrier in the electron transfer or in other processes like mass transport and coupled chemical reactions.

The potential difference ε_0 in an electrochemical cell in the equilibrium state is constant. In contrast, as soon as a current flows through an electrochemical cell, the potential is shifted depending on the direction and intensity of the flowing current. This effect will be the topic of this chapter.

In Fig. 2.24(a) and (b), the shift of the cell potential in a galvanic cell and in an electrolytic cell is shown.

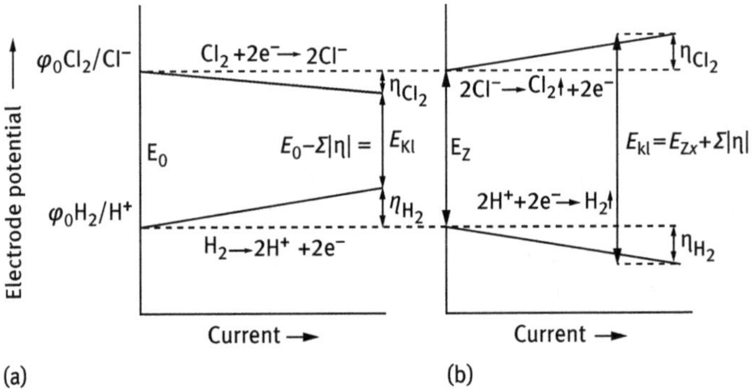

Fig. 2.24: Schematic representation of the course of the overvoltage in the galvanic cell H_2/HCl (a) and of the terminal potential difference in the HCl electrolytic cell (b).

Remember, there exists an activation energy for the transfer of charge across the interface.

2.5.3 Charge transfer and activation energy

Now some questions appear: Which are the components of energy forming this barrier? and "How is the heights of this barrier influenced by the potential of the electrode?" Figures 2.25 and 2.26 give answers to these questions. Figure 2.25 a schematic illustration of the energies involved in the transfer of electrons across the electrode/electrolyte interface at a redox electrode. It must be emphasized that no potential from outside is put on the metal. It is just the Galvani potential in the interior of the

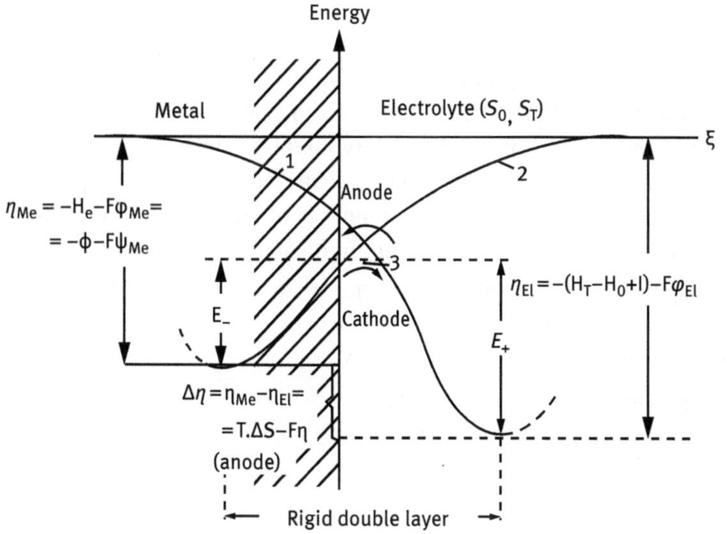

Fig. 2.25: Schematic representation of the energies that influence the electron transfer in a redox reaction.

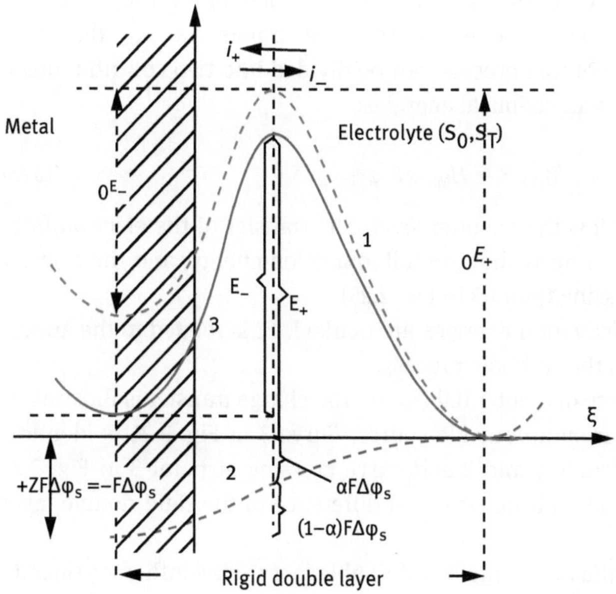

Fig. 2.26: The dependence of the anodic activation energies E_+ and the cathodic activation energy E_- on the potential difference $\Delta\varphi_r$ in the rigid double layer.

metal that determines the energy level of the electron in the metal. Some further comments that are helpful for understanding are given in Figure 2.25.[15]

The energy E of the electron is plotted versus the distance ξ. At the crossing of the axis is the origin of energy, that is, the energy of the electron in the charge-free vacuum. The origin of the ξ-axis is in the boundary metal/electrolyte, that is, it intersects the rigid double layer very close to the metal surface.

Curve 1 represents the course of the potential energy of an electron that is transferred from the redox system S_O/S_R to the charge-free vacuum. Dividing its energy h_{el} into partial energies involved in the process $S_R \longrightarrow S_O + e^-$ (vacuum) are listed below; hence, we come to the bottom of the problem.

These energies are listed as follows:

Desolvation of S_r to $S_r{}^*$	$+ H_r$
Ionization of $S_r{}^* \rightleftharpoons S_o{}^*$	$+ I_r$
Solvation of $S_o{}^*$ to S_o	$- H_o$
Overcoming of the Galvani potential φ	$+ F\varphi$

Thus, the energy h_{el} of the electron in the solvated redox system comes to

$$h_{el} = -H_r + H_o - I - F\varphi \quad (2.126)$$

Curve 2 represents the transfer of the electron from the vacuum level into the interior of the metal phase, that is, to the level of the Galvani potential φ_{Me}, that is, to the Fermi level. The energy of this process can be divided into two contributions $F\varphi_{Me}$ and H_{Me} which includes all chemical energies:

$$h_{Me} = -H_{Me} - F\varphi_{Me} \quad (2.127)$$

Curves 1 and 2 intersect below the vacuum level. The transfer of the electron from the electrolyte into the metal needs not the full ionization energy, and the transfer is possible at the curve crossing (point 3 in Fig. 2.25).

In Fig. 2.25, different activation energies are marked. E_+ is related to the anodic process and E_- is related to the cathodic process.

In Fig. 2.26, the influence of a potential φ_{Me} on the charge transfer is illustrated; it is applied on the metal from an outside source. Curve 3 in Fig. 2.26 is identical with the superposition of curve 1 and 2 and curve crossing at point 3 in Fig. 2.25 illustrating the transfer reaction if no potential difference in the rigid double layer exists.

Besides the surface of the metal, the "rigid double layer" may influence the rate of charge transfer.

15 The energies refer to 1 mol of electrons = $6.02 \cdot 10^{23}$ electrons.

Applying from an outside source, the potential φ_{Me} leads to the potential difference $\Delta\varphi_r = -\varphi_{el} > 0$ in the rigid double layer (curve 2). The superposition of curves 1 and 2 leads to curve 3. Comparison of curve 1 and curve 3 reveals that the activation energies depend on the potential φ_{Me}. The activation energy E_+ is smaller than ${}_0E_+$, whereas the activation energy E_- is enhanced in comparison to ${}_0E_-$.

If the transfer of an electron from the electrolyte to the metal is faster, the more positive the metal is charged with respect to the electrolyte and vice versa. A cathodic process will be faster with more negative potential of the metal, whereas the anodic process will be slower. Changes of the transfer rates with changes of the potential cause changes in the current. Thus, it is obvious that the current depends on the potential.

Curve 1 may be considered as the "chemical energy curve" that omits electrical interactions. Curve 3 may be termed the "electrochemical" energy curve. The comparison of both curves reveals the influence of the electrode potential. The anodic activation energy of the "electrochemical" energy curve is lowered by $\alpha F\,\Delta\varphi_r$ and the cathodic activation energy is enhanced by $(1-\alpha)F\,\Delta\varphi_r$. The transmission factor α[16] marks the position of the maximum of the activation barrier in the rigid double layer. It depends not on the potential and has the value $0 < \alpha < 1$; generally it is taken to be very close to 0.5.

The dependence of the activation energies E_+ and E_- from the potential is described by

$$E_+ = {}_0E_+ - \alpha F\,\Delta\varphi_r \tag{2.128}$$

$$E_- = {}_0E_-(1-\alpha)F\,\Delta\varphi_{r\,-} \tag{2.129}$$

It should be noticed that the decay of the potential in the rigid double layer, as shown in Fig. 2.26 is a good approximation for solutions with high concentrations of an electrolyte. This is realized in most of the experiments where a supporting electrolyte is used. If this is not the case, one adds to the potential difference $\Delta\varphi_r$ the potential difference $\Delta\varphi_{dl}$ occurring in the diffuse double layer.

Above a qualitative description of the electrochemical kinetics is given as a foretaste of the following quantitative description. The final goal is, of course, a potential–current relation.

16 T. Erdey-Gruz, M. Volmer, Z. Phys. Chem. 150A, 203 (1930).

2.5.4 Comparison of chemical and electrochemical kinetics

The kinetic world is not an isolated world; it is connected to the world of thermodynamics. Perhaps, it may be easier for the reader to start the following section by showing this relation.

Consider two substances A and B linked by a mobile equilibrium, for example, a cis–trans rearrangement:

$$A \rightleftharpoons B \tag{2.130}$$

Mobile equilibrium means the forward and the backward reactions are active at all times. Considering a unimolecular reaction the rate of the forward process is v_f (mol/s):

$$v_f = k_f c_A \tag{2.131}$$

the rate of the reverse reaction is

$$v_b = k_b c_B \tag{2.132}$$

The rate constants k_f and k_b have the dimension s^{-1}. The net conversion rate v_{net} is

$$v_{net} = k_f c_A - k_b c_B \tag{2.133}$$

At equilibrium the net conversion rate is zero; hence,

$$k_f / k_b = K = c_B / c_A \tag{2.134}$$

The kinetic theory predicts a constant concentration ratio at equilibrium just as thermodynamics does.

Such an agreement is required for any kinetic theory. In the limits of equilibrium, kinetic equations must collapse to relations of thermodynamics; otherwise, the kinetic picture cannot be accurate.

In the equilibrium state, the rates of the rearrangement of A to B and vice versa are equal, which leads to the definition of the exchange velocity v_0:

$$v_0 = k_f (c_A)_{eq} = k_b (c_B)_{eq} \tag{2.135}$$

The exchange velocity plays an important role in the treatment of electrode kinetics.

Most of the rate constants of solution reactions follow a linear relation between $\ln k$ and $1/T$. Arrhenius was the first who recognized the generality of this relation and proposed the formulation

$$k = A \exp - E_a / RT \tag{2.136}$$

nowadays well known as Arrhenius relation.

E_a has the dimension of an energy and is well known as the activation energy. The exponential factor is reminiscent of the probability of using thermal energy to surmount an energy barrier of height E_a. A is called the frequency factor (sometimes the A factor). A must be related to the frequency of attempts to surmount the energy barrier E_a. May be, this picture is an oversimplification, but it carries the essence of truth and is useful for casting a mental image how reactions proceed.

The charge transfer between metal and depolarizer in the electrolyte follows the same law. As shown in Fig. 2.26, an activation barrier exists for the charge transfer across the boundary metal/electrolyte. However, in contrast to a chemical reaction in an electrochemical reaction the activation energies E_+ and E_- depend on the potential applied to the electrode from an outside voltage source. Shifting the potential in the positive direction favors the anodic current, and shifting it to the negative direction favors the cathodic current.

At the equilibrium potential E_e where both currents cancel each other, the electrode is in its equilibrium state. The equilibrium state of an electrochemical reaction

$$S_O + \mathrm{ne} \longleftrightarrow S_R \tag{2.137}$$

is described by the Nernst equation

$$E = E_O{}' + RT/nF\ln(c_O{}^*/c_R{}^*) \tag{2.138}$$

where $c_O{}^*$ and $c_R{}^*$ are the bulk concentrations and $E_{O'}$ is called normal potential. , if NHE is the reference electrode.

For corresponding conditions, any theory of electrode kinetics must predict this result.

Early investigations of the current close to the equilibrium revealed that the observed current is strictly controlled by the interfacial transfer dynamics. Near the equilibrium potential the current is connected exponentially with the overvoltage:

$$I = a' \exp \eta/b' \tag{2.139}$$

or as formulated by Tafel[17]

$$\eta = a + b \log I \tag{2.140}$$

Equation (2.140) is known as the Tafel equation.

In the following, the foundation of this empirical formula will be given.

2.5.5 Derivation of the current–overpotential equation

The current–overpotential relation is the core of electrode kinetics and shall now be explained.

17 J. Tafel, Z. Phys. Chem., 50, 641 (1905).

Like the chemical equilibria the redox equilibrium (2.134) is a mobile equilibrium, which means the observed current I consists of a forward current i_f and a backward current i_b or more illustrative it involves a cathodic current i_c and an anodic current i_a. The observed current i is therefore

$$I = i_a + i_c \tag{2.141}$$

The cathodic reaction may be considered as the forward reaction. It proceeds at a rate v_f, which must be proportional to the surface concentration of the component S_O. The constant of proportionality k_f links the rate to the concentration $c_O(x,t)$; it is termed rate constant k_f:

$$v_f = k_f c_O(x,t) = i_c/nFA \tag{2.142}$$

The concentration is a function of the distance x from the surface and the time t. Hence, the surface concentration is $c_O(0,t)$. Since the forward reaction is a reduction the current is a (negative) cathodic current. In analogy we formulate the backward reaction, which is an oxidation

$$v_b = k_b c_R(x,t) = i_a/nFA \tag{2.143}$$

where i_a is the anodic component of the total current. The resulting reaction rate is

$$v_{net} = v_f - v_b = k_f\, c_O(0,t) - k_b\, c_R(0,t) = i/nFA \tag{2.144}$$

or for the current i

$$i = i_f - i_b = nFA(k_f\, c_O(0,t) - k_b\, c_R(0,t)) \tag{2.145}$$

It has to be noticed that reaction velocities in heterogeneous systems refer to unit interfacial area and hence have the dimension $mols^{-1}\ cm^{-2}$. If the concentrations are expressed in mol/cm^3 the dimension of the rate constant is cm/s.

The concentrations in eqs. (2.144) and (2.145) refer to surface species participating in the heterogeneous redox reaction. Their kinetic behavior is strongly influenced by the interfacial potential difference.

Reactions at the surface of the electrode can be controlled by the applied potential $(E - E_O)$. Two equations predict the dependence of k_f and k_b on the difference $(E - E_O)$:

$$k_f = k_O \exp(-n\,\alpha\, f\,(E - E_O)) \tag{2.146}$$

$$k_b = k_O \exp(-n(1-\alpha) f (E - - E_O)) \tag{2.147}$$

where $f = F/RT$ and k_O and α are adjustable parameters called the standard rate constant and the transfer coefficient. Equations (2.146) and (2.147) resemble the Arrhenius equation.

Substituting relations (2.146) and (2.147) into eq. (2.145) yields the complete current–potential equation:

$$I = nFA\ k_O\{c_O(0,t)\exp(-n\ \alpha f(E-E^0)) - c_R(0,t)\exp(-n(1-\alpha)f(E-E^0))\} \quad (2.148)$$

This relation is very important because modifications of this equation are used in the treatment of every problem connected with heterogeneous kinetics.

k_O is a measure of the kinetic facility of a redox couple. A system with a large values of k_O will achieve equilibrium in a short time, but systems with a small k_O will be sluggish. The largest values of standard rate constants are in the range between 1 and 10 cm/s. They are associated with simple electron transfer processes, like the reduction or oxidation of aromatic hydrocarbons, where the geometry of the reactants and products is very similar. The process needs only electron transfer and negligibly small rearrangement of the solvation shell.

The transfer coefficient α is a measure of the symmetry of the energy barrier. In most systems, its value lies between 0.3 and 0.7; usually it is approximated by 0.5.[18]

At equilibrium, the velocity of the forward and backward reactions is identical, and the net current is zero. We require the electrode to adopt a potential based on bulk concentrations of O and R as dictated by the Nernst equation. From eq. (2.144) follows for the case $i = 0$:

$$nFA\ k_O(c_O(0,t)\exp(-n\alpha\ F/RT(E-E_O)) = nFA\ k_O c_R(0,t)\exp(-n(1-\alpha)\ F/RT(E-E_O)) \quad (2.149)$$

At equilibrium, the bulk concentration $c_o{}^*$ of S_O and $c_R{}^*$ of S_R are also found at the surface; hence

$$\exp(nF/RT(E_{eq}-E_O)) = c_O{}^*/c_R{}^* \quad (2.150)$$

which is simply the exponential form of the Nernst equation.

The balanced activity can be expressed in terms of the exchange current i_O, which is equal in magnitude to either component i_c or i_a.

If both sides of eq. (2.149) are raised to the $-\alpha$ power, we obtain

$$\exp(-\alpha n\ F/RT(E_{eq}-E_O)) = c_O{}^*/c_R{}^{*-\alpha} \quad (2.151)$$

Substitution of eq. (2.150) into eq. (2.149) results in

$$i_O = nFA\ k_O c_O{}^*\exp(1-\alpha)c_R{}^{*\alpha} \quad (2.152)$$

The exchange current I_O is proportional to k_O. It can often be substituted for k_O in kinetic equations. For the special case $c_O{}^* = c_R{}^* = c$ the exchange current is given by

18 A.J. Bard, F. L. Faulkner, "Electrochemical Methods", J. Wiley & Sons 1980, p. 96ff.

$$I_0 = nFA\ k_0\ c \tag{2.153}$$

In many publications, the exchange current is normalized to unit area of the electrode, which defines the exchange current density *j*; $j = i_0/A$.

The current *i* can be expressed by using i_0. The resulting equation is the current–overpotential equation:

$$i = i_0(c_o(0,t)/c_o)^*\ \exp(-\alpha nF\eta/RT) - c_R(0,t)/c_R{}^*\exp((1-\alpha)nF\eta/RT) \tag{2.154}$$

In the following we have to make some remarks on the current–overpotential equation.

2.5.6 The Butler–Volmer equation

The current–overpotential equation can be simplified, provided the solution is well stirred or currents are kept so low, so that the surface concentrations do not differ appreciably from the bulk values. Under these conditions, one gets the *Butler–Volmer equation*:

$$i = i_0(\exp - \alpha n\ F\eta/RT) - \exp((1-\alpha)nF\eta/RT) \tag{2.155}$$

This equation is a good approximation of eq. (2.154) if the current *i* is less than about 10% of the limiting current i_a or i_c.

In Fig. 2.27 (Bard[18] p. 102), the current overpotential curve of a redox system is shown. It consists of the cathodic and anodic components.

The overpotential η associated with any given current images an activation energy. The lower the exchange current the more sluggish are the kinetics; hence, the larger is the activation overpotential for any particular net current. If the exchange current is very large (case a in Fig. 2.28), then the system can supply large currents with insignificant activation overpotential of charge transfer at the electrode. If in such a case overpotential is observed it is concentration overpotential because the transport of electroactive species from the bulk of solution to the electrode surface becomes the rate-determining step of the overall process.

When mass transfer effects begin to manifest themselves, then the concentration overpotential will contribute, but the bulk of overpotential is for the activation of charge transfer. In this kind of system, the reduction wave occurs at much more negative potentials, and the oxidation wave lies at much more positive values than E_0. Such systems are termed quasireversible or irreversible.

The shape of the current-overpotential curve also depends on the transfer coefficient α (fig. 2.29). The exchange current can be viewed as a kind of "idle speed" for charge exchange across the interface. If we want to draw a net current that is only a small fraction of this bidirectional idle current, then a very small overpotential will be required to extract it. Even at equilibrium, the system is delivering charge across the

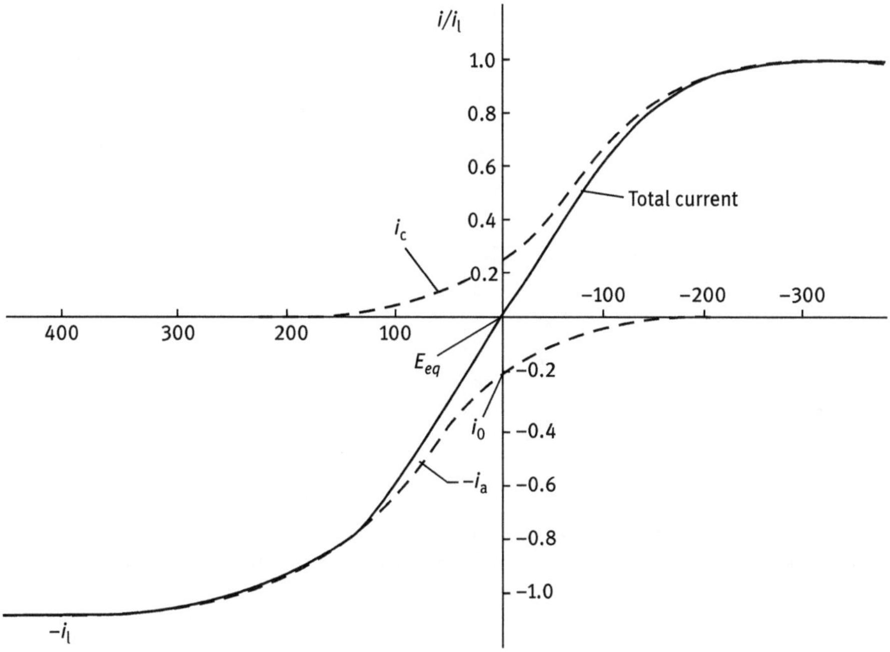

Fig. 2.27: Current–potential curves for the system O + $e^{-} \rightleftharpoons$ R with $\alpha = 0.5$, $n = 1$, $T = 298$ K, the limiting currents $i_{l,c} = i_{l,a} = i_l$. The dotted lines show the component currents i_a and i_c.

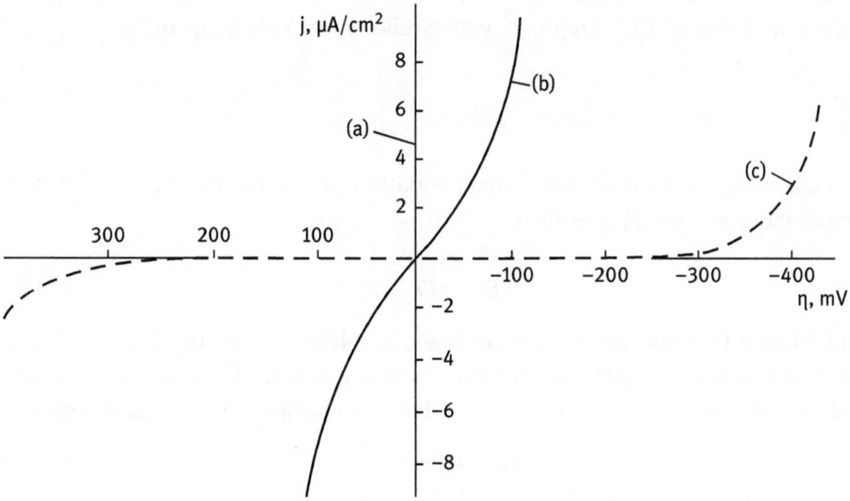

Fig. 2.28: Effect of the exchange current density on the activation potential required to deliver net current densities. (a) $j_{0-} = 10^{-3}$ A/cm^2, (b) $j_0 = 10^{-6}$ A/cm^2, (c) $j_0 = 10^{-9}$ A/cm^2. For all cases, the reaction is O + ne $\rightleftharpoons$ R with -$\alpha = 0.5$, $n = 1$ and $T = 298$ K.

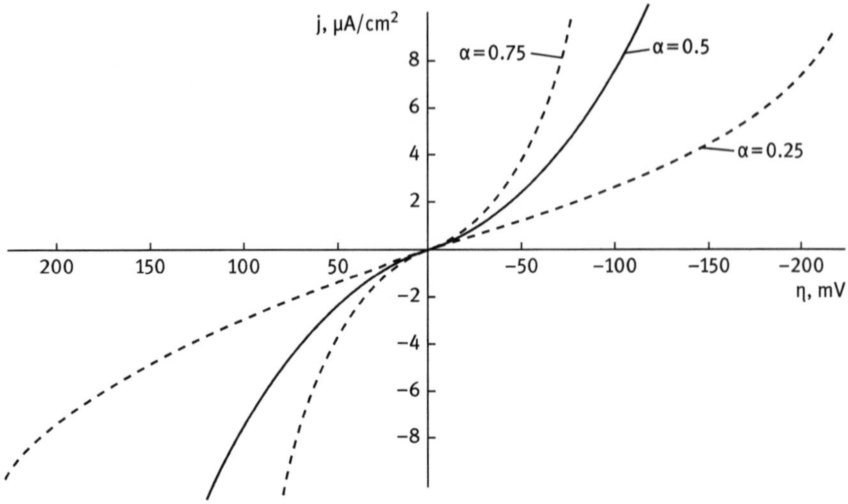

Fig. 2.29: Effect of the transfer coefficient on the symmetry of the current–overpotential curves for O + ne $\rightleftharpoons$ R with $n = 1$, $T = 298$ K and $j_0 = 10^{-6}$ A/cm^2 here mass transfer is not included.

interface at rates much greater than we require. The small overpotential unbalances to small degree the rates in the two directions so that one of them predominates. On the other hand, if we ask for a net current that exceeds the exchange current, we have to apply significant overpotential.

Exchange currents in real systems are observed in the wide range between 10 amp/cm^2 and several 10^{-12} amp/cm^2, which reflects the wide range of k_0.

2.5.7 The Tafel behavior of the overpotential

For small values of η the Butler–Volmer equation can be linearized, and e^x can be approximated as $1 + x$. This leads to

$$i = i_0(-nF\eta/RT) \tag{2.156}$$

which shows that the net current is linearly related to the overpotential η in a narrow potential range near the equilibrium potential E_0. The ratio η/R has the dimension of a resistance and is called the charge transfer resistance R_{ct}:

$$R_{ct} = RT/nF\ i_0 \tag{2.157}$$

In some experiments R_{ct} can be evaluated directly. It serves as a convenient index of kinetic facility; for large k_0 it clearly approaches zero.

For large values of overpotential we observe the Tafel behavior. In this case, one of the bracketed terms (either positive or negative) in the Butler–Volmer equation becomes negligible. For example, at large negative overpotentials $\exp(-\alpha n F\eta/RT)$ $\exp((1-\alpha)nF\eta/RT)$, and the Butler–Volmer equation

$$i = i_0 \exp(-\alpha n\ F\eta/RT) - \exp((1-\alpha)nF\eta/RT)$$

becomes

$$i = i_0 \exp(-\alpha n\ F\ \eta/RT) \tag{2.158}$$

or

$$\eta = RT/\alpha n\ F\ \ln i_0 - RT/\alpha n\ F\ \ln i \tag{2.159}$$

This relation corresponds to the empirical Tafel equation (2.140), and the empirical Tafel constants a and b can now be identified from eq. (2.159):

$$a = -2.3RT/\alpha n\ F\ \log i_0 = -0.0591/\alpha n\ \log i_0 (25\,°\mathrm{C}) \tag{2.160a}$$

$$b = -\ 2.3RT/\alpha n\ F = -0.0591/\alpha n\ (25\,°\mathrm{C}) \tag{2.160b}$$

The Tafel form can be expected to hold whenever the backreaction contributes less than 1% of the current, which implies at 25 °C that

$$|\eta| > 0.118/\mathrm{nV} \tag{2.161}$$

An experimentalist searching for the first time the linear Tafel relation may be disappointed, because he cannot find it. The reason is, if the electrode kinetics are fairly facile, one will approach the mass transfer-limited current, that is, the kinetics of the total process is controlled partly by mass transport from the bulk of the solution to the surface of the electrode (see below). When the electrode kinetics are sluggish and significant, overpotentials are required, and the good Tafel relationships can be seen. This shows that the Tafel behavior is an indicator for totally irreversible kinetics. Systems in that category allow no significant current flow except at high overpotentials where the Faraday process is effectively unidirectional and therefore irreversible.

Figure 2.30 shows the "ideal" Tafel plots.

Tafel plots are useful for evaluating kinetic parameters. They have an anodic branch (slope: $(1-\alpha)\ nF/2.3\ RT$) and a cathodic branch (slope: $-\alpha nF/2.3\ RT$). The linear segments can be extrapolated to the intercept where $\log i_0$ is detected. The transfer coefficient α and the exchange current i_0 can be taken from this kind of presentation. However, depending on the nature of the system under investigation, there are often limitations by the influence of mass transfer. This leads to deviations from the linear section, which may hide the linear section completely. As already mentioned, charge transfer through the interface and mass transfer from the bulk of solution to the surface are linked sequentially in every system. If the rate of mass transfer is smaller than the rate of charge transfer, the complete process is controlled by mass transfer.

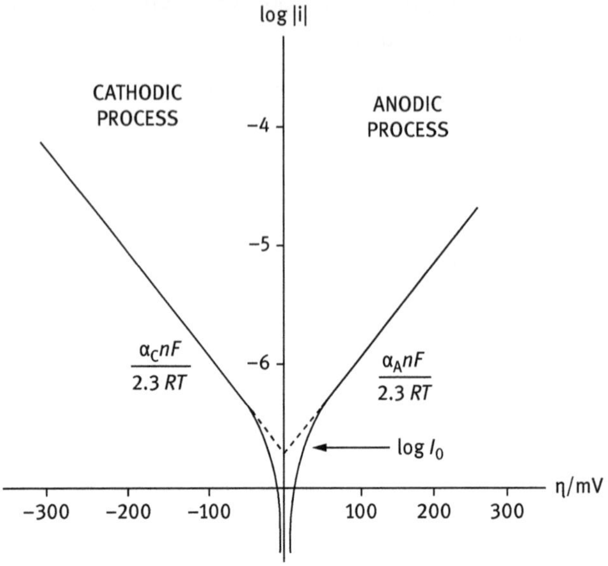

Fig. 2.30: Tafel plots or anodic and cathodic linear branches of the current–overpotential curve for $S_O + ne^- \rightleftharpoons S_R$ with $n = 1$, $\alpha = 0.5$ and $j_O = 10^{-6}$ amp/s.

As an example, in Fig. 2.31 Tafel plots measured in the redox couple Mn(IV)/Mn (III) are shown.[19]

2.5.8 Reversible and irreversible electrode processes

What means "reversible behavior" of an electrochemical system?

We mentioned earlier that there are systems where no kinetic parameters can be observed. This obviously is the case when i_O, respectively k_O, are very large. We rewrite the current–potential curve (2.153) as follows:

$$i/i_O = (c_o(0,t)/c_o{}^*)\exp(-\alpha nF\eta/RT) - (c_R(0,t)/c_R{}^*)\exp((1-\alpha)nF\eta/RT \qquad (2.162)$$

When i_O becomes very large in comparison to the current of interest, the ratio i/i_O approaches zero and we can rearrange eq. (2.162) to

$$c_o(0,t/c_R(0,t) = c_o{}^*\exp(nF/RT)/c_R{}^* \qquad (2.163)$$

and by substitution from the Nernst equation in the form

$$c_o(0,t/c_R(0,t) = \exp(nF/RT)(E-E_O) \qquad (2.164)$$

19 K.J. Vetter, G. Manecke, Z. Phys. Chem (Leipzig) 195, 337 (1950).

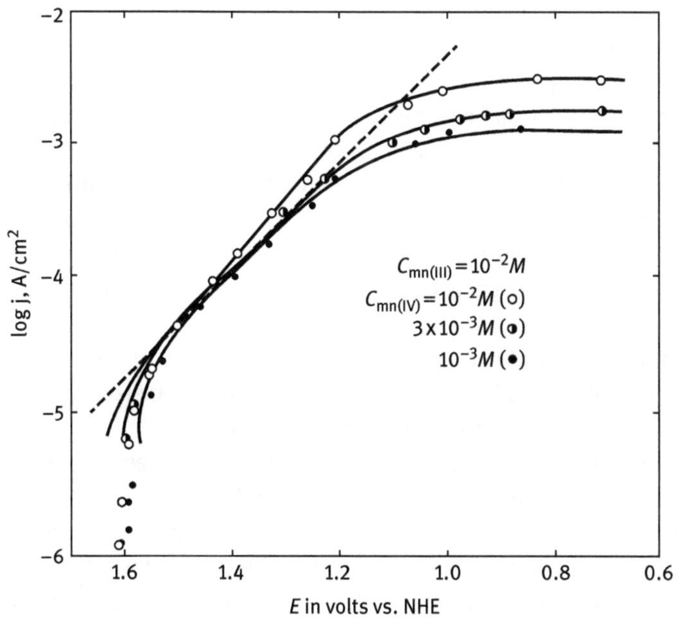

Fig. 2.31: Tafel plot from the reduction of Mn(IV) to Mn(III) at Pt in 7.5 M H_2SO_4 at 298 K. The dashed line corresponds to $\alpha = 0.24$.

we get the very important equation

$$E = E_0 + RT/nF \ln(c_o(0, t/c_R(0, t)) \tag{2.165}$$

We see that the electrode potential and the surface concentration of S_O and S_R regardless of the current flow are linked by an equation of the Nernst form.

If the activation barrier of charge transfer is very low, no experimental manifestations of kinetic charge transfer parameters can be seen. By the fast charge transfer the equilibrium between the electrode potential and the surface concentration is adjusted, so that the thermodynamic equation (2.165) always holds. However, the surface concentration and the bulk concentration of the electroactive species are not in equilibrium. Therefore, the net current is controlled by the flow of material from the bulk of solution to the surface. In unstirred solutions i_o is diffusion controlled.

2.6 Mass transport

2.6.1 Diffusion, migration and convection

In the preceding section we discussed the charge transfer across the electrode/electrolyte boundary, and it has been pointed out that charge transfer and mass transport are inseparably coupled together. This justifies a deeper inspection of mass transport in electrochemical cells. The key parameter of mass transport is the gradient of the electrochemical potential.

The concentration of the electroactive species changes at the electrode surface. This causes a gradient in concentration. Transport by diffusion of the electroactive educts of the electrochemical reaction from the bulk of solution to the surface of the electrode and of the products back to the bulk of the electrolyte counteracts this gradient. Charge exchange occurring at the surface of electrodes changes the concentrations of the electroactive species. This process is fixed in space and surrounded by the bulk of the solution.

In a simple system consisting of two electrodes in an electrolyte during an electrolysis, a potential difference exists between the electrodes. Near the electrode surface and in the bulk of solution exist gradients of the potential.

Consequently, charged particles migrate in the gradients of the electric field. In the bulk of solution, gradients of the concentration hardly exist.

If a gradient in the electrochemical potential μ_e exists, mass transport occurs by diffusion, by migration or by convection. Depending on the experimental conditions, these transport mechanisms dominate. Stirred and unstirred solutions make a difference in the transport mechanism and differences appear whether ions or neutral molecules are considered.

2.6.2 A general view

Consider a solution having species i with two points r and s in a small distance from each other with different electrochemical potentials, $\mu_e^r \neq \mu_e^s$.

The difference $\Delta\mu_e^i$ is a gradient of the electrochemical potential arising from different concentrations (respectively, activities) of species i at r and s (a concentration gradient) or because of a difference in the potential ϕ (an electric field or potential gradient). In general, a flux of species i will occur to equalize this difference of μ_e^i.

The following discussion is restricted to the mass transfer in x-direction, which is perpendicular to the planar electrode in a stirred solution. For flux J of species i in x-direction, the following equation is used:

$$J_i = -(c_i D_i / R\ T)\ \delta\mu_e^i / \delta x \tag{2.166}$$

The electrochemical potential μ_e^i is given by

$$\mu_e^i = \mu_=^i + RT \ln a_i + z_i F\ \phi \tag{2.167}$$

The gradient $\delta\mu_e^j/\delta x$ of the electrochemical potential contains three contributions due to diffusion, migration and convection. This is expressed by the Nernst–Planck equation (2.168), which shows the different contribution to the flux J_i:

$$J_i(x) = -D_i\delta c_i(x)/\delta x - z_i F/R\,T \cdot D_i c_i \delta\phi(x)/\delta x + c_i v(x) \tag{2.168}$$

The first term describes the contribution of diffusion, the second term describes the contribution of migration and the third takes into account the contribution of convection. The term $c_i\,v(x)$ is zero in unstirred solutions.

For describing the mass transport in unstirred solutions remains

$$J_i(x) = -D_i\delta c_i(x)/\delta x - z_i\,F/R\,T \cdot D_i c_i \delta\phi(x)/\delta x \tag{2.169}$$

As in eq. (2.169), concentrations may be introduced instead of activities.

The negative sign of the first term results from the fact that the concentration $c_i(0)$ of the species i at the electrode surface is always smaller than the concentration in the bulk of the solution. This means $\mu_e i$ increases in x-direction but the flux of species i occurs in $-x$ direction, that is from the bulk of solution to the electrode. The sign of the second term depends on the sign of the potential and of the ions moving in the electrical field. For positively charged ions moving to a negatively charged electrode, we get a negative sign of the second term in eq. (2.169).

For the y- and z-directions in an orthogonal system exist equations analogous to eq. (2.169).

The flux J is equivalent to current i if the species i are ions. The flux J_i (mol/scm^2) is equal to $-i_i/z_iF\,A$(Cb/s/mol cm^2), where i_i is the current component at any value of x arising from the flow of species i.

The flux J_i or the current i_i can be divided into the diffusion current i_d and the migration current i_m:

$$-J_i = i_i/z_iF\,A = i_{d,\,i}/z_iF\,A + i_{m,\,i}/z_iF\,A \tag{2.170}$$

Of course, the question arises: How much each of the two components contributes to the total flux? The answer will be given in the following section.

2.6.3 The contribution of diffusion and of migration to the mass transport

The relative contribution of diffusion and migration to the flux of a species and the contribution of the flux of that species to the total current differ at a given time for different locations in solution. Near the electrode, the electroactive substance i is, in general, transported by both processes. The flux of this substance at the electrode surface controls the rate of the electrode reaction.

We use a system that already has been described in Section 2.3 in conjunction with Hittorf transfer numbers to find out the ratio i_d/i_m of the diffusion current i_d to the migration current i_m. It is the electrolysis of a solution of hydrogen chloride at platinum electrodes.

We assume that 10 e per unit time pass the cell producing 5 mol of H_2 at the cathode and 5 mol of Cl_2 at the anode. The transfer numbers are $t^+ = 0.8$ and $t^- = 0.2$ and the ratio of the specific conductivity of H^+ to Cl^- is 4:1. According to the different mobility 8 H^+ migrate per unit time toward the cathode and 2 Cl^- migrate toward the anode. In order to maintain the steady current and electroneutrality, however, 10 H^+ must be supplied per unit time. This means that an additional 2 H^+ diffuse to the electrode, bringing along 2 Cl^- to maintain electroneutrality. Similarly at the anode to supply 10 Cl^- per unit time, in addition 8 Cl^- must be supplied by diffusion along with 8 H^+. Thus, the different currents (in arbitrary e-units) are: for H^+, $i_d = 2$, $i_m = 8$; for Cl^-, $i_d = 8$, $i_m = 2$. The total current is 10. In this case, migration and diffusion point in the same direction.

In mixtures of charged species, the fraction of the total current i carried by cations is t_i, and the amount carried by anions is $t_i \cdot i$. The number of moles of the ith species migrating per second is $t_i\, i/z_j\, F\, i$. If the ith species is undergoing electrolysis, the moles electrolyzed per second are $|i/n\, F$, while the moles arriving at the electrode per second by migration are $\pm i_m/n\, F$, where the positive sign applies to reduction of i and the negative sign pertains to oxidation. Thus

$$\pm\, i_m/n\, F = t_i i/z_i F \quad i_m = \pm\, n/z_i \cdot t_i\, i \tag{2.171}$$

and since

$$i_d = i - i_m \quad i_d = i(1 \pm n\; t_i/z_i) \tag{2.172}$$

where the minus sign is used for cathodic currents and the positive sign for anodic currents: $i_{d,i}$ and $i_{m,i}$ are diffusion and migration currents of species i, respectively.

The factor $(z_i\, F\, D_i/R\, T)$ is the mobility u_i of species i as described in Section 2.3, so that

$$i_{m,i}/|z_i|F\, A = u_i c_i \delta\phi/\delta x \tag{2.173}$$

At any location in solution during electrolysis, the total current i is made up of contributions from all species, that is

$$I = \sum{}^{i} i_i \tag{2.174}$$

The current for each species results from the diffusional component arising from a concentration gradient and the migrational component arising from a potential gradient.

The relative contribution of diffusion and migration to the flux of a species and of the flux of that species to the total current differs at a given time for different locations in solution. Near the electrode the electroactive substance i is, in general, transported by both processes. The flux of this substance at the electrode surface controls the rate of the electrode reaction and, therefore, the faradaic current is flowing in the external circuit. As shown earlier that current can be separated into diffusion and migration currents reflecting the diffusive and migrational components to the flux of the electroactive species at the surface:

$$i = i_d + i_m \tag{2.175}$$

In the bulk of a solution (away from the electrodes), concentration gradients are generally small, and the total current is carried mainly by migration. All charged species contribute. For j in the bulk region of a linear mass transfer system having a cross-sectional area A, $i_i = i_{m,i}$ or

$$I_i = |z_i| F\ A\ u_i c_i\ \delta\phi/\delta x \tag{2.176}$$

For a linear electric field

$$\delta\phi/\delta x = \Delta E/l$$

where $\Delta E/l$ is the gradient (V/cm), arising from the change in potential ΔE over distance l. Thus,

$$i_i = |z_i| F\ A\ u_i c_i \cdot \Delta E/l \tag{2.177}$$

From eq. (2.174) the total current in the bulk of the solution is given by

$$I = \sum\nolimits^{i} i_i = F\ A \Delta E/l \cdot \sum\nolimits^{i} |z_i| u_i c_i \tag{2.178}$$

While migration carries the current in the bulk of the solution during electrolysis in the vicinity of the electrodes, concentration gradients of the electroactive species arise and diffusional transport also occurs. Indeed under some circumstances, as we will see, the total flux of electroactive species to the electrode is due to diffusion.

2.6.4 The role of supporting electrolytes

In many electrochemical experiments, supporting electrolytes are used. Supporting electrolytes consist of ions, which are electroinactive in a wide potential range. Typical ions of supporting electrolytes are alkali ions and tetraalkylammonium ions and the anions F^-, ClO_4^-, PF_6^-, BF_4^-.

Supporting electrolytes are used in concentrations of about 0.1 mol/L; in comparison the concentration of an electroactive depolarizer is in the range 10^{-3}–10^{-4} mol/L. Supporting electrolytes do not participate in the charge transfer at the electrodes. However, because their concentration is high they dominate the migration current in the bulk of the solution. The analysis of the transport mechanisms of cell containing supporting electrolyte shows that most of the electroactive species now reach the cathode or anode by diffusion. Only less than 0.5% of the total Faraday current is due to migration. In most of the electrochemical investigation, supporting electrolytes are used, which means that under this condition diffusion is the dominating mechanism for mass transport.

The effect of supporting electrolyte on the limiting current can be illustrated by the results compiled in Table 2.11 showing the limiting current for the reduction of Pb(II) at the dropping mercury electrode (polarography) with different amounts of KNO_3 added to the solution. At low concentrations of KNO_3, Pb(II) carries an appreciable fraction of the total current. At higher KNO_3 concentrations, K^+ migration replaces the contribution of the Pb(II) species, and the observed limiting current caused by the reduction of Pb^{2+} is essentially a pure diffusion current.[20]

Table 2.11: The effect of added KNO_3 on the limiting current due to reduction of Pb(II) (the concentration of $PbCl_2$ was $0.95 \cdot 10^{-3}$ M).

Added KNO_3 (M)	0	0.0001	0.0002	0.0005	0.001	0.005	0.1	1.0
Limiting current (μA)	17.6	16.2	15.0	13.4	12.0	9.8	8.45	8.45

Another important function of the supporting electrolyte is a decrease in cell resistance (and in analytical applications it may decrease or eliminate matrix effects). Especially in cells with nonaqueous electrolytes (organic solvents), the addition of supporting electrolyte is a must.

A final remark: For the practical use, very high purity of supporting electrolytes is important and – of course – the “open range” of the supporting electrolyte must be adapted to the potential range where an electrode process is investigated.

2.6.5 Diffusion

Since in most of the electrochemical experiments an excess of supporting electrolyte is used, the mass transfer by diffusion plays the most important role in such experiments. It is therefore appropriate to deal with the phenomenon of diffusion in more detail.

2.6.5.1 Fick’s laws of diffusion

On a macroscopic scale, diffusion is described by the two laws of Fick.

20 J.J. Lingane and I.M. Kolthoff, J. Am. Chem. Soc., 61, 1045 (1939).

If excess of supporting electrolyte is used in a reasonable approximation, the migration of electroactive species may be neglected. Kinetics and mechanism of electrode processes are strongly influenced by diffusion.

In the following, we will consider two common experimental situations: experiments in an unstirred solution and experiments in a stirred solution represented by the rotating disk electrode.

2.6.5.2 Linear diffusion to a planar electrode in an unstirred solution

Diffusion is the only form of effective mass transport of electroactive species to an electrode in an unstirred electrolyte containing an excess of electroinactive supporting electrolyte.

We consider the transport of the oxidized species O and the reduced species R connected by the redox reaction

$$O + ne^- \longrightarrow R^- \tag{2.179}$$

Figure 2.32 illustrates the flux of the species O and R. O moves in the direction perpendicular to the plane of the electrode. The produced R at the electrode moves in the opposite direction into the bulk of solution. The flux of the electrons is directed into the electrode.

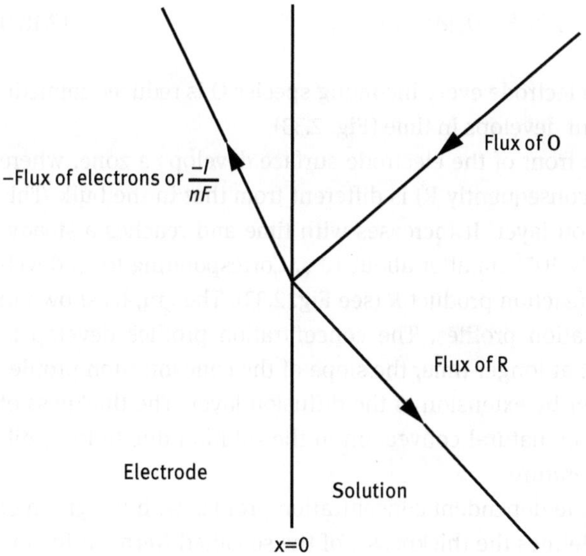

Fig. 2.32: The flux balance at the electrode surface drawn for the reaction $O + ne^- \rightarrow R^-$.

According to the first law of Fick, the flux of the species O and R is given by the relation

$$-\,\text{flux}(\text{O}) = D_\text{O}\partial c_\text{O}/\partial x \tag{2.180}$$

and

$$\text{flux}(\text{R}) = D_\text{R}\partial c_\text{R}/\partial x \tag{2.181}$$

The flux is defined by the number of molecules (respectively, moles) passing perpendicular per second the unit area (cm^2). The negative sign takes into account the difference between the direction of movement and concentration gradient. D_O and D_R are the diffusion coefficients of O and R. Typically values of the diffusion coefficient are of the order of 10^{-5} cm^2/s. Relations (2.180) and (2.181) represent Fick's first law: The flux of a species perpendicular through the unit area is proportional to the concentration gradient of this species.

Assuming that the charge transfer across the boundary electrolyte/electrode is very fast, every species O reaching the electrode surface is immediately reduced to R. Thus, we can formulate a relation between the flux of O and R and the flux of electrons across the boundary electrode/solution, that is, the current I:

$$I/n\,F = -\,D_\text{O}[\partial c_\text{O}/\partial x]_{x=0} \tag{2.182}$$

and

$$I/n\,F = D_\text{R}[\partial c_\text{R}/\partial x]_{x=0} \tag{2.183}$$

If due to the potential of the electrode every incoming species O is reduced immediately, a concentration gradient develops in time (Fig. 2.33).

As shown in Fig. 2.33, in front of the electrode surface develops a zone, where the concentration of O (and consequently R) is different from that in the bulk. This zone is known as the diffusion layer. It increases with time and reaches a steady-state value of approximatively 10^{-2} cm after about 10 s. Corresponding to c_O develops the concentration of the reaction product R (see Fig. 2.33). The graphs shown in Fig 2.33 are called concentration profiles. The concentration profiles develop in time. As $t \to 0$ they are steep; at longer time, the slope of the concentration profiles becomes continuously smaller by extension of the diffusion layer. The thickness of the diffusion layer is limited by natural convection in the solution due to local differences in density and temperature.

The intersection of the time-dependent concentration profiles with the graph of the bulk concentration c_O^∞ defines the thickness x of the so-called Nernst diffusion layer.

The closer examination of the concentration profiles reveals that the concentration c_O is a function of the distance x and the time t: $c_O = f(x,t)$; an analogous relation exists for c_R.

Equation (2.184) is known as the second law of Fick.

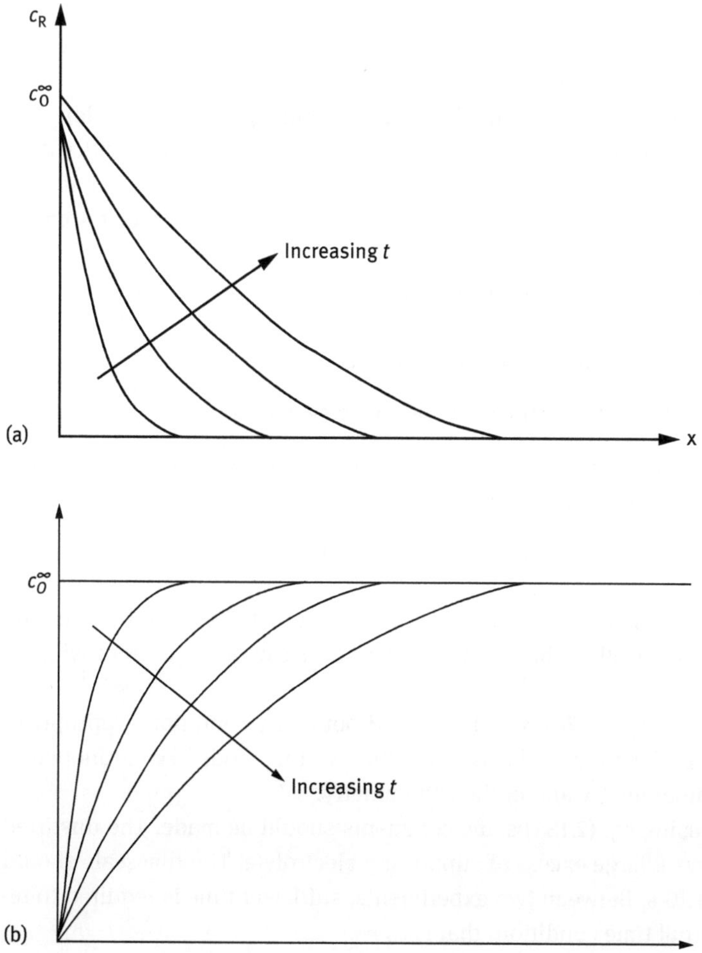

Fig. 2.33: Time evolution of the concentration profile of O and R of reaction $O + ne^- \longrightarrow R^-$ at a potential where the process is diffusion controlled. At $t = 0$ for all x is $c_= = c_=^*$ and for $t > 0$ is at $x = 0$ $c_= = 0$ and at $x = \infty$ is $c_= = c_=^*$. $c_=^*$ is the concentration in the bulk of the solution.

$$\partial c_O/\partial t = D_O\, \partial^2 c_O/\partial x^2 \qquad (2.184)$$

It says: The development of the concentration in time on a given space coordinate x is proportional to the second derivative of the concentration to the square of this coordinate. With increasing time, diffusion will cause the concentration profiles to relax to their steady state by extending into the solution and becoming less steep. The flux of O to the electrode decreases and thus the current density also decreases in time (Fig. 2.33).

For the solution of eq. (2.184), one needs the initial and boundary conditions of the experiment under consideration. The mathematical procedure to solve Fick's second law is described in Appendix G.

Figure 2.33 illustrates the solution of eq. (2.184), that is, the time development of the concentration profiles of O and R during an experiment carried out with a solution initially containing O but not R. The electrode potential is stepped at $t = 0$ in such a way that the surface concentration of O changes instantaneously from the bulk concentration c_O^∞ to zero.

Thus the boundary conditions of the experiment are

$$\text{at } t = 0 \text{ for all } x, \; c_O = c_O^\infty$$

$$\text{as } t > 0 \; (\text{at } x = 0) \; c_O = 0 \text{ and at } x = \infty, \; c_O = c_O^\infty$$

The form of the $I - t$ transient, that is, the solution of Ficks' second law under the boundary condition of the problem is given by

$$I = (n\,F\,D^{1/2} c_O^\infty)/(\pi^{1/2} t^{1/2}) \tag{2.185}$$

As shown in Fig. 2.33, the current time response at a potential where the electrode process is diffusion controlled shows an exponential decay of the current with increasing time.

The integration of eq. (2.184) with initial and boundary conditions appropriate to the particular experiment is the basis of the theory of methods like chronopotentiometry, chronoamperometry and cyclic voltammetry.

To be honest, using eq. (2.184) some comments should be made. The unstirred solution should have a large excess of supporting electrolyte. The timescale should be limited to about 10 s. Between two experiments, sufficient time is required to restore the correct initial time condition, that is $c_O = c_O^\infty$.

2.6.5.3 Mass transport in a stirred solution

The rotating disk electrode

The rotating disk electrode is one of the mostly used tools of electrochemistry. It consists (Fig. 2.34) of a small metal electrode (diameter 1–2 mm) encapsulated into a glass capillary. The radius of the surrounding glass is about ten times that of the electrode. The surface of the electrode and the glass capillary is carefully polished. The rotation frequency ω of the electrode can be precisely measured.

The rotating electrode sucks solution up to the disk spinning it out across the edge of the glass sheath. Hence, the concentration changes at the electrode surface are influenced by both the diffusion and the convection. The term convectional diffusion is used for this kind of mass transport.

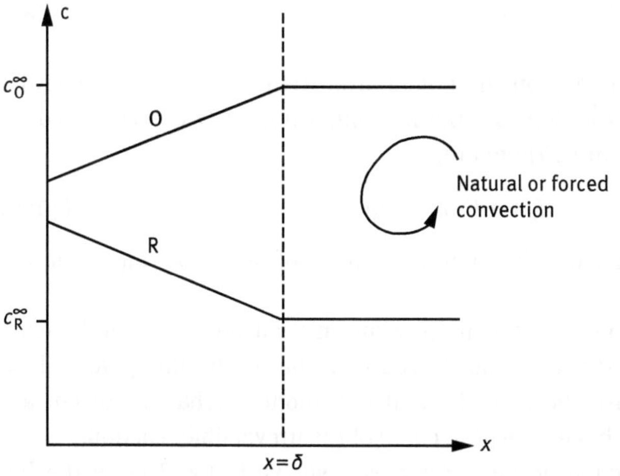

Fig. 2.34: Steady-state profile for the reaction O + ne⁻ → R˙. The solution contains $c_O{}^\infty = 3\ c_R{}^\infty$.

For the dependence of the concentration c_i on the time t, the following equation is valid:

$$\Delta c_i/\delta t = D_i \delta^2 c_i/\delta x^2 + V_X \delta c_i \delta x \quad (2.186)$$

The first term of this equation relates to diffusion, and the second term to convection. V_x is the speed of the solution perpendicular to the disk. For the solution of eq. (2.186), the relationship between V_x and the rotation rate ω on the distance x perpendicular to the disk has to be known. Levich[21] deduced the following equation:

$$V_x = \omega^{3/2} x^2 \quad (2.187)$$

which shows that the importance of convection increases with the square of the distance x from the electrode surface. This allows the definition of a boundary layer of the thickness δ, near the electrode surface in which convection perpendicular to the electrode surface becomes zero and therefore diffusion is the only significant form of mass transport. Outside this boundary layer, convection maintains the concentration of all species uniform and at their bulk values. A steady-state profile of a solution containing O and R is shown in Fig. 2.34.

The thickness of the diffusion layer is determined by the rotation rate of the electrode. It decreases with increasing rotation rate. In the steady state, the plot c_i versus x in a reasonable approximation is taken to be linear. The steady-state current I is given as follows:

21 B. Levich, "Physicochemical Hydrodynamics", Prentice Hall, 1962.

$$I = -n\,F\,D\,(dc_O/dx)_{x=0} = \mathrm{n}F\,D\,(c_O{}^{\infty} - c_O{}^{s})/\delta \qquad (2.188)$$

The concentration of O depends on the potential of the electrode. At a potential where $c_O{}^s = 0$, at a given rotation rate the maximum flux of O is reached corresponding to a limiting current I_L. Therefore,

$$I_L = n\,F\,D\,c_O{}^{\infty}/\delta \qquad (2.189)$$

reaches a plateau. This indicates the limiting currents of reduction and oxidation given by eq. (2.189).

In the potential range between the limiting currents, mixed control of the reaction by electron transfer and mass transfer occurs. At the equilibrium potential E_e, the current is zero, because the cathodic and the anodic exchange current are equal. Notice the different shape of the *I–E* curve of the irreversible reaction.

A more detailed inspection of the *I–E* curves is shown in fig. 2.35. In the *I–E* curve of the irreversible reaction close to the equilibrium potential E_e the regions a, where, due to the small current (Fig. 2.35b) we can conclude that the extend of the corresponding chemical change at the electrode surface is small; hence, the surface concentrations of O and R are not significantly different from the bulk concentrations. Thus, the influence of mass transport on the electrode process is small and the small current density I (<5% of I_d) is controlled by the electron transfer. It is this range *a* where Tafel plots can be measured. The other limiting situation is labelled by *c*. In this range, due to the large overpotential the electron transfer is very fast and the electrode process is controlled by the mass transfer from the bulk to the electrode surface. In case of the rotating disk electrode this is given by the flux in the diffusion zone (eq. (2.189)). In the range labeled b, we have mixed control of the electrode process by charge transfer and by mass transfer.

The *I–E* curve of the reversible electrode process shows neither mixed control nor electron transfer control of the electrode reaction. The reason is that the electron transfer is very fast, that is, both the anodic and the cathodic exchange current are very high, and therefore O and R at the electrode surface are always in equilibrium. It is the case where thermodynamic arguments, in fact the Nernst equation, can be applied for calculating the surface concentrations. Of course, if the electron transfer is very fast, no overpotential appears and the electrode process is controlled by the mass transport.

The *E–I* curve of the reversible system (Fig. 2.35a) can be described by an equation based on the Nernst equation

$$E = E^0 + RT/nF\ \ln c_s(\mathrm{O})/c_s(\mathrm{R}) \qquad (2.190)$$

and by equations 2.191 and 2.192 for the anodic and the cathodic current derived by Levich:

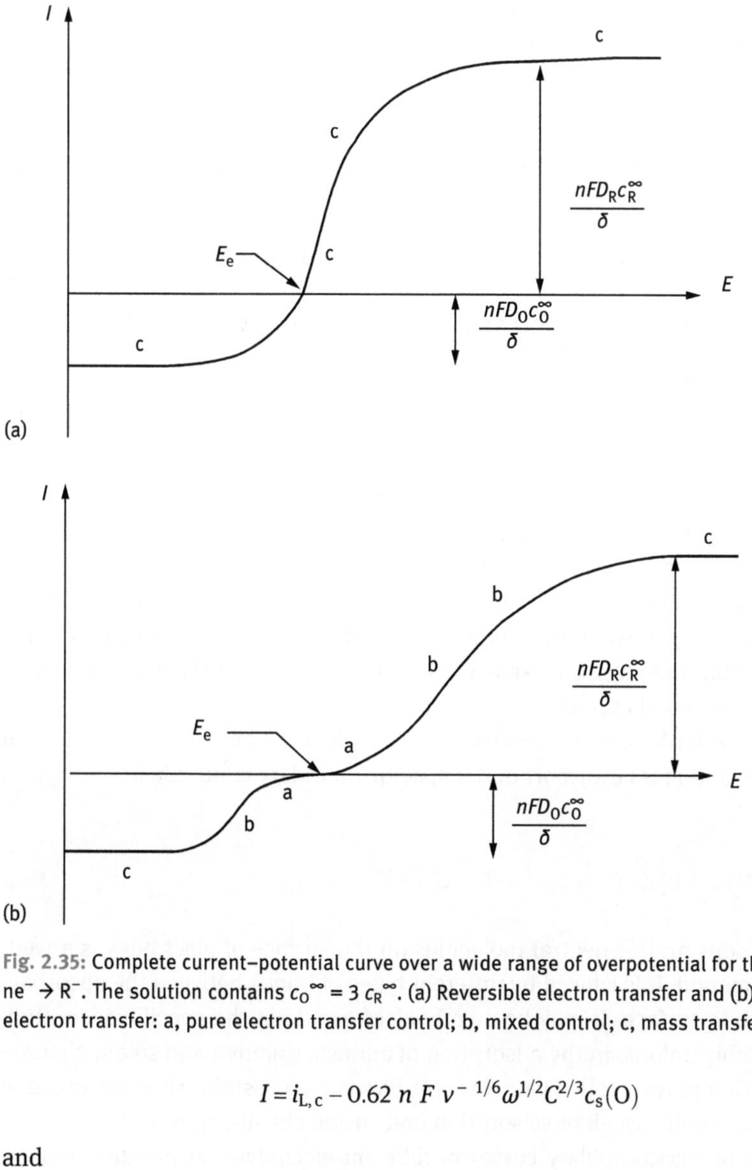

Fig. 2.35: Complete current–potential curve over a wide range of overpotential for the reaction O + ne⁻ → R⁻. The solution contains $c_O^\infty = 3\, c_R^\infty$. (a) Reversible electron transfer and (b) irreversible electron transfer: a, pure electron transfer control; b, mixed control; c, mass transfer control.

$$I = i_{L,c} - 0.62\, n\, F\, \nu^{-1/6}\omega^{1/2}C^{2/3}c_s(O) \tag{2.191}$$

and

$$I = i_{L,a} + 0.62\, n\, F\, \nu^{-1/6}\omega^{1/2}D_R^{2/3}c_s(R) \tag{2.192}$$

where D_O and D_R are the diffusion coefficients of O and R, respectively; $i_{L,a}$ and $i_{L,c}$ are the anodic and the cathodic limiting current.

Substituting eqs. (2.191) and (2.192) into the Nernst equation yields

$$E = E^0 + RT/nF\ \ln D_R^{2/3}/D_o^{2/3} + RT/nF \ln(i_{L,c} - i)/(i - i_{L,a}) \tag{2.193}$$

This equation may be reduced to eq. (2.194) for the anodic current if the solution contains only the reduced depolarizer R:

$$E = E^0 + RT/nF \ln D_R{}^{2/3}/D_o{}^{2/3} - RT/nF \ln(i_{L,a} - i)/i \qquad (2.194)$$

and in analogy to the equation of the cathodic current, if the solution contains only the oxidized depolarizer O

$$E = E^0 + RT/nF \ \ln D_R{}^{2/3}/D_o{}^{2/3} + RT/nF \ln(i_{L,c} - i)/I \qquad (2.195)$$

Since the diffusion coefficients of O and R are very similar, eqs. (2.194) and (2.195) can be simplified to

$$E = E^0 \pm RT/nF \ln(i_L - i)/i \qquad (2.196)$$

Conclusion

For the presentation given earlier, we have chosen the simplest electrochemical system in which the charge transfer across the surface metal/electrolyte is coupled with the mass transfer of the electroactive species O to the electrode surface. This is a sequential coupling of two kinetic processes. Since in steady state the mass transport is slower than the electron transfer across the electrode surface, we observe a mass transfer-controlled current.

In one of the forthcoming chapters (3.4) we will come back to the coupling of charge transfer and mass transport under specific boundary conditions.

2.7 Adsorption and phase formation

Adsorption of ions and/or neutral molecules on the surface of electrodes is a well-known phenomenon. It is caused by different types of bonds between the adsorbate and the electrode surface. It may be merely electrostatic in the specific adsorption of ions, preferably anions. In the adsorption of amines, thiourea and so on, charge–dipole interaction plays a role. Even covalent bonds are possible. Thus the range of bond energies contains weak physisorption and strong chemisorption.

The shape of electrocapillary curves of different electrolytes at mercury was the first hint that for the formation of the electric double layer not only the long-range electrostatic interaction is involved but also other short-range forces. The electrocapillary curves show different appearance at positive potentials; the positions of their maxima obviously depend on the nature of the anions. When the adsorbate is an electroactive species the electrode reaction may be strongly affected, and species not involved in the electrode process may change the rate of charge transfer and the final product.

The extent of adsorption is generally expressed by the surface coverage Θ, which is defined by the fraction of the surface covered by the adsorbate. Θ depends on the electrolyte, the solvent, adsorbate structure and adsorbate concentration and on the concentration of any other species in the system as well as the nature of the electrode surface. Sometimes the surface coverage changes with the electrode potential. A positively charged electrode attracts preferably anions, while a negatively charged electrode promotes the adsorption of cations. Charge at the electrode surface interacts with dipole molecules leading to their adsorption and orientation at the surface. At the PZC the uncharged electrode, preferably the adsorption of neutral molecules occurs.

The coverage of an adsorbate is related to its concentration in solution by an isotherm:

$$\Theta/(1-\Theta) = c^{\infty} \exp(-\Delta G_{\mathrm{ads}}/RT) \tag{2.197}$$

This is the Langmuir isotherm. It is the simplest isotherm. Lateral interactions of the adsorbate molecules are neglected, that is, this isotherm describes a two-dimensional analogon to the "ideal gas." There are several other isotherms, which consider the lateral interaction. One of them is the Frumkin isotherm. It assumes that the free energy increases linearly with the coverage:

$$\Delta G_{\mathrm{ads}} = \Delta G^{0}{}_{\mathrm{ads}} + g\,\Theta \tag{2.198}$$

Leading to the isotherm

$$\Theta/(1-\Theta) \exp\ g\Theta/RT = c^{\infty} \exp\ (-\Delta G^{0}{}_{\mathrm{ads}}/RT) \tag{2.199}$$

Isotherms are usually deduced from the measurement of capacitance as a function of the bulk concentration c^{∞} and the potential E.

The adsorption of an electroinactive species can affect the kinetics of a redox couple in solution by reducing the rate. There are two reasons for reducing this rate. The area or the electron transfer can be reduced by adsorption or the electron transfer must occur over a larger distance.

Adsorption of the electroactive species itself makes electron transfer difficult, while adsorption of the product makes electron transfer easier.

Adsorption of intermediates, whether generated by electron transfer, for example

$$\mathrm{H^{+} + e^{-} \longrightarrow H_{ads}} \tag{2.200}$$

or by predissociation of a bond in the electroactive species, for example

$$\mathrm{CH_3OH \longrightarrow H_{ads} + CH_2OH_{ads}} \tag{2.201}$$

are important in electrocatalysis.

The degree of adsorption increases with time, and the surface concentration of the adsorbate increases. Aggregation of the adsorbed molecules leads to the formation of a critical germ that has the structure of a new phase. This process is called nucleation. The nuclei grow forming a two-dimensional phase.

Generally, the lattice of this phase is not commensurable with the lattice at the electrode surface. As an example, a thymine film on gold <111> is shown in Fig. 2.36.

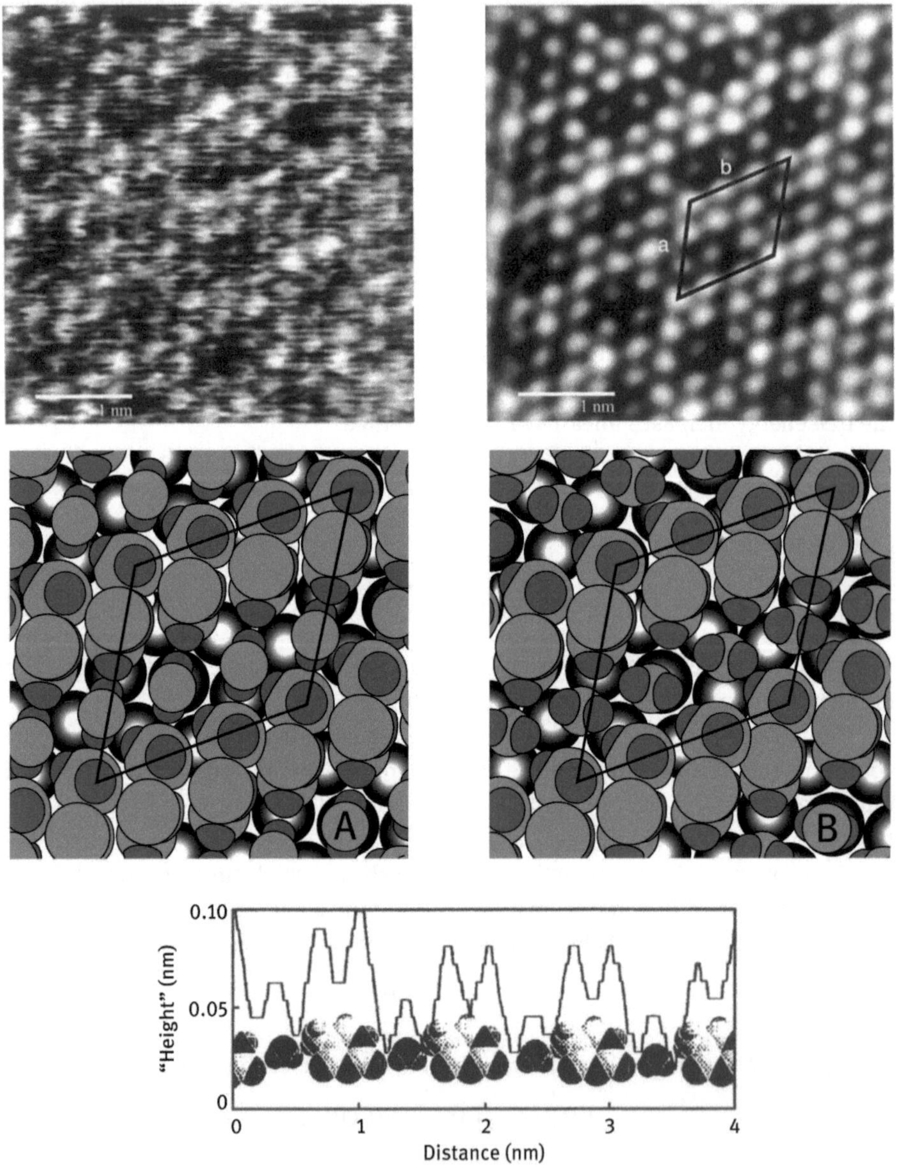

Fig. 2.36: In this figure a thymin film on is Au<111> shown. It has been obtained by an in situ STM measurement. The upper left figure shows the unfiltered measurement. In the upper right figure the signal/noise ratio has been improved by filtering. The lower figures A and B show models of the film. The thymine molecules form aggregates, their plane is perpendicular oriented versus the gold plane; thymine forms linear aggregates which are connected by water molecules.

The formation of a film in a nucleation growth process is governed by the intermolecular interaction of the adsorbed molecules; in aqueous systems water molecules at the surface may be involved in the structure of the film. The formation of films and their stability can be investigated by measuring capacity/potential curves. One observes pits with low capacity, and the potential range of the pit shows the range of stability of a film and at the pit edges the kinetics of film formation and film disintegration (Fig. 2.37).

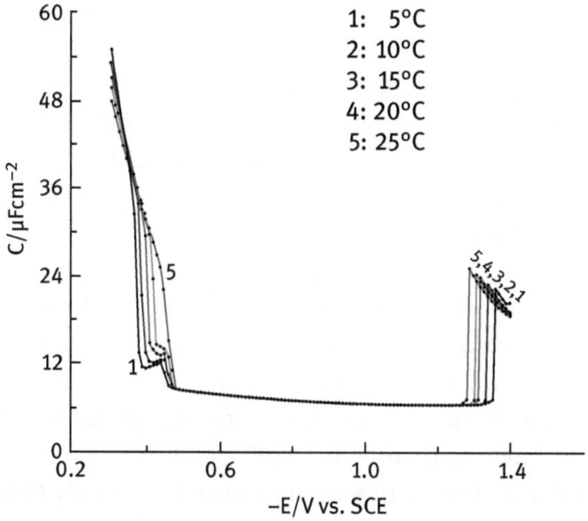

Fig. 2.37: The pit in the differential capacity/voltage curve due to filmformation of 6-n-propyl-2 thiouracil in a $3 \cdot 10^{-3}$ mol/l aqueous solution at different temperatures. By detailed analysis of the pitedge one gets information on the kinetics of filmformation.

Another type of film is formed by the sulfur-containing molecules, for example, thiourea. The adsorbed molecules are arranged to a surface lattice by binding the sulfur atom in a threefold hollow site of the gold <111> lattice. Thus, the lattice of the adsorbate is commensurable with the gold lattice. This kind of adsorbate is very stable.

It suggests itself that adsorption and film formation change the property of an electrode; in most cases, the electron transfer will be more difficult, only when the product of a reaction is adsorbed the electron transfer may be facilitated.

2.8 Solid electrodes and modern electrochemistry

2.8.1 The electrode/electrolyte interface

Most of the early investigations of the double layer have been performed by using mercury electrodes, which have a smooth and reproducible surface. The first insight into the nature of the phase boundary metal/electrolyte results from investigating the electrocapillarity at mercury electrodes.

Since the very early days of electrochemistry, however, solid metal electrodes have been used widely in research as well as in industrial applications. A great number of metals have been used for cathodes. Pb, Cd, Sn, Fe, Ni, Cr, Mo and so on are examples. For the choice of a cathode the hydrogen overvoltage of the metal is often important. The higher the hydrogen overpotential the higher working potentials are possible for a cathodic reaction in aqueous medium.

Noble metals such as Au, Pt and Pd play a special role. For performing redox processes preferably the noble metals, especially platinum, have been used. They have a reasonable range of stability against oxidation. In addition there are suitable metal oxide electrodes. Lead dioxide is a well-known material used as anode material. Graphite, artificial coal and glassy carbon are used as anode materials.

Up to now preferably polycrystalline materials have been used as electrodes for the variety of different electrochemical experiments.

Electrodes are prepared by different specific procedures, which often include mechanical polishing and cycling of the potential in the range of stability of the electrode. The quality of the electrode is tested by using well-known electrochemical reactions.

The surface of polycrystalline electrodes consists of a mixture of grains with different crystallographic planes at the surface. Thus, measuring the properties of these electrodes one can expect only averaged properties of the surface.

Studying the double layer of solid electrodes suffered in former days not only from the difficulty to prepare clean, reproducible surfaces of electrodes but also from the lacking knowledge of the surface structure.. For a long time nearly nothing was known about the structure of the boundary solid electrode/electrolyte. Anyhow, it has been shown that the double layer capacity of the electrode depends on the metal used.

The situation changed when single crystal electrodes could be prepared and handled by experimental techniques developed in surface science. The preponderance has been on noble metal surfaces. There is an emphasis on studies employing exactly the same experimental probes used to study the solid/gas surfaces, that is by LEED (low-energy electron diffraction), GIXS (grazing incidence X-ray scattering), SEXAFS (surface-extended X-ray adsorption fine structure) and STM (scanning tunneling microscopy).

Nowadays, a lot of experimental experience exists for preparing clean surfaces of solid single crystal electrodes. The modern techniques have been described in recent literature of surface science[22] and electrochemistry.[23]

22 K. Christmann, "Introduction to Surface Physical Chemistry", Springer Verlag, New York, 1991, p. 83ff.

23 P.N. Ross in "Structure of Electrified Interfaces", VCH Publishers, New York 1993 p. 35ff.

For the preparation of clean and ordered single crystal surfaces (fig. 2.38), three methods are known.

In the mostly used method, single crystal rods or boules, grown by zone refining, are oriented by the Laue back-reflection technique and then cut by spark erosion methods along the desired crystal face. The near-surface layers rendered amorphous by the cutting process are removed by chemical dissolution. Subsequently, metallographic polishing is done with successively finer grades of alumina or diamond paste to obtain a uniformly smooth surface.

The second method is based on the fact that spherical single crystals are obtained when a polycrystalline Au wire is melted in a gas-oxygene flame. The heat treatment may also serve to purify the surface from trace surface and bulk impurities. The single crystal beads are then oriented, cut and metallographically polished. The oriented and polished rods are reannealed at near-melting temperatures to repair the damaged selvage.

The third procedure, limited to <111> faces, involves epitaxial growth by vapor deposition onto a mica substrate. This method is most expedient for the preparation of Pt<111> single crystal, thin-film electrodes such as those employed in the study of Langmuir–Blodgett films and self-assembled monolayers, since the material, relative to the platinum metals, is more resistant to surface contamination and roughening; pretreatments to restore the surface order cleanliness, which could be detrimental to the adhesion of the Au film on the substrate, would not be required.[24]

Nowadays, single crystal electrodes are available commercially.

One principal problem of electrochemistry is the knowledge of the structure of the electrode/electrolyte interface. To make a good progress, it was necessary to study the surface of single crystal electrodes in contact with the electrolyte. Determining the structure of surfaces in contact with the solution may be divided into two categories, so-called ex situ methods, where the electrode is removed from the electrolyte under controlled potential and under ultrahigh vacuum conditions is transferred into a suitable equipment for further analytical investigation. Using this technique one assumes that the double layer still exists at the wet surface. In the alternative “in situ” method the electrode surface is probed by various spectroscopic methods in the electrolyte and under potential control.

Many different electrodes with atomically flat and well-defined surfaces have been investigated. It turned out that many close-packed metal surfaces are thermodynamically stable, but many other metal surfaces are not. Resulting from the imbalance of forces at the surface small changes of the bond lengths normal to the surface may occur. Typically, a change in the order of a few percent of the bulk spacing is observed between the first and second layers of metal atoms followed by a dilatation between the second and third layers. This is called surface relaxation.

---

24 M.P. Soriaga in “Structure of Electrified Interfaces”, VCH Weinheim, 1993, p. 106.

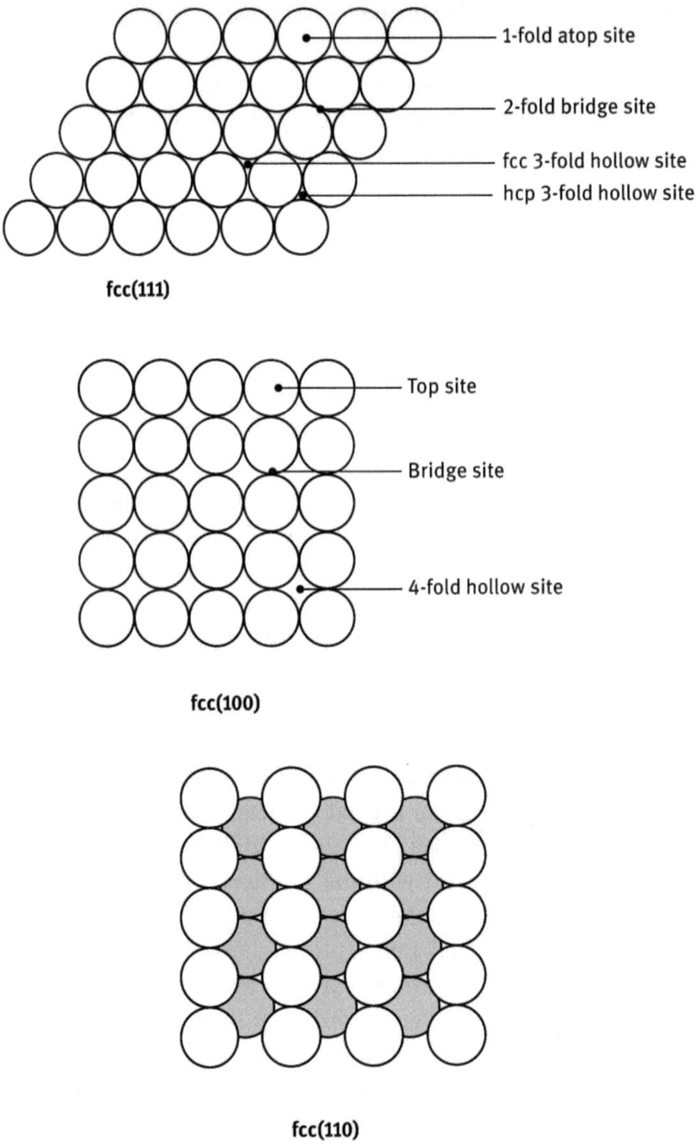

Fig. 2.38: Top view of the atomic arrangement in the <111>, <110>,<100> crystallographic planes of a face-centered (fcc) cubic crystal.

Larger scale changes, facilitated by surface diffusion over distances much greater than atomic spacing, are called surface restructuring. It may cause macroscopic changes in the surface topography, like creation of stakes and kinks, faceting and surface roughening.

These structural effects are relevant for interfacial electrochemistry, but also for kinetics of electron transfer across the boundary metal/electrolyte or for the deposition of metals as performed in surface improvement. The influence of structural effects has been documented ever since single crystal electrodes are investigated.[25,26,27]

Another effect is surface reconstruction. This is a lateral displacement of surface atoms. It can be connected with phase transitions at the surface.

The question appears: Can reconstruction be observed at single crystal electrodes in contact with the electrolyte?

Considering the results of very detailed experiments with single crystal electrodes of noble metals, this question can be answered. Gold electrodes[28] are a characteristic example on which we restrict the answer to the question (fig. 2.39).

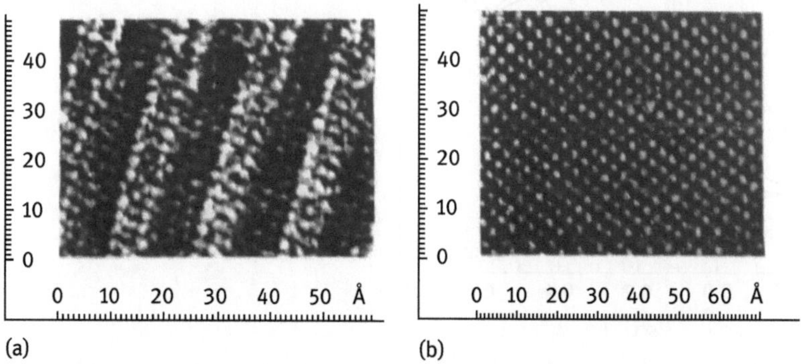

Fig. 2.39: STM images of the reconstructed and unreconstructed gold surface: (a) Au<100>–(5 × 20) at $U = -0.24$ V versus SCE (reconstructed) and (b) Au<100>–(1 × 1) at $U = +0.56$ V versus SCE in 0.1 M H_2SO_4 (unreconstructed). The reconstructed surface (bright lines in fig 2.6.2) is slightly buckled with a periodicity of 14.5 Å and a height of about 0.7–0.8 Å.

This example demonstrates convincingly the experimental properties available for studying the surface of solid electrodes. Besides electrochemical methods, STM and atomic force microscopy, many other methods originally used in surface science have greatly enhanced our knowledge about detailed mechanisms of surface structural changes and their influence on the double layer.

Reconstructed gold surfaces are stable in contact with aqueous electrolytes, if certain potential requirements are met. The Au(100)–(5 × 20) reconstructed electrode is stable in the potential range between −0.35 and +0.25 V (vs SCE); however,

25 A. Hamelin in "Modern Aspects of Electrochemistry", Plenum Press, New York 1985, Vol. 16, p. 1ff.

26 P.N. Ross, F.T. Wagner in "Advances in Electrochemistry and Electrochemical Engineering", Wiley New York, 1984, Vol. 13, p. 69ff.

27 D.M. Kolb, Ber. Bunsenges. Phys. Chem. 1988, 92, 1175.

28 D.M. Kolb in "Structure of Electrified Interfaces", VCH Publishers, New York, 1993, p. 65ff.

a single potential excursion to +0.70 V removes the reconstruction and the Au(100) surface acquires the (1 × 1) form. Specific adsorption of anions as well as underpotential deposition of Cu removes the reconstruction. Figure 2.40 shows a capacity–potential curve of the double layer.

The potential of zero charge (PZC) of the reconstructed and the unreconstructed surface is remarkably different (table 2.12). In current–potential curves, the different capacities give rise to sharp spikes caused by loading the double

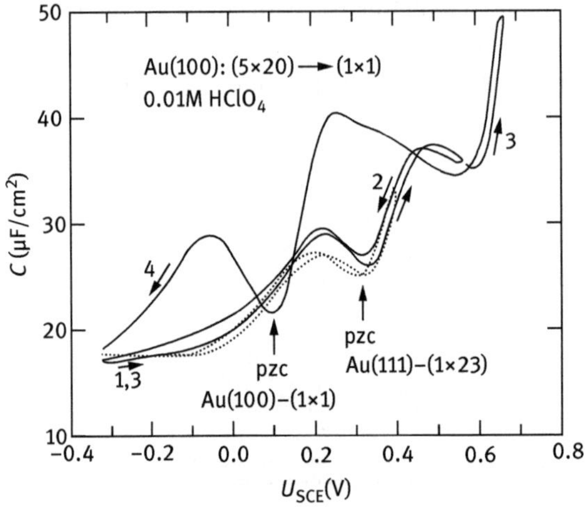

Fig. 2.40: Double layer capacity–potential curve for Au(100)– (5 × 20) in 0.01 M $HClO_4$ with transition from (5 × 20) to (1 × 1). Numbers refer to a sequence of scans; dots indicate the capacity curve for Au (111)–(1 × 23).

Table 2.12: Potentials of zero charge for reconstructed and unreconstructed gold single-crystal electrodes in perchloric acid solution[29].

Surface structure	PZC (V vs SCE)
Au(100)–(5×20)	+0.30
Au(100)–(1 × 1)	+0.08
Au(111)–(1 × 23)	+0.32
Au(111)–(1 × 1)	+0.23
Au(110)–(1 × 2)	≈ −0.04
Au(110)–(1 × 1)	−0.02

29 D.M. Kolb, J. Schneider, J. Surf. Sci. 1985, 162, 764.

layer. Provided no charge transfer through the phase boundary occurs, they may be observed if slow scan rates (≤50 mV/s) are used.

2.8.2 Electric conductivity of solids

The electrochemical cell consists of an electrolyte and electrodes, mainly metal electrodes but also semiconductor electrodes. So far liquid electrolytes have been in the focus of interest where the transport of electric charge occurs due to the mobility of solvated ions. Now a short presentation of the electric conductivity in solids, that is, metals, semiconductors and ionic crystals, will be given.

The experimental key parameters are the specific resistance ρ or the conductivity κ of the solid, which depend on the mobility of the electric charge. As shown in Table 2.1, they can be very different for different materials. The electrical conductivity of solids is in the range $10^6/\Omega/cm$ for metals down to $10^{-18}/\Omega/cm$ for quartz; isolators are found in the range of 10^{-6}–$10^{-18}/\Omega/cm$.[30]

In the following we will focus on semiconductors because they are used in semiconductor electrochemistry which is important for light harvesting to gain electric energy. Semiconductor electrochemistry has forced the alliance of solid state physics and electrochemistry. The convention for representing the semiconductor/electrolyte interface is largely based on the energy band diagram used in solid state physics.

2.8.3 Electrons in solids

There are different models for describing electrons in solids. We start with a simple model on the electric conductivity of electrons in metals.

The electron gas

The free-moving electrons in a metal are termed the electron gas. It describes the electrons like a classical gas, taking not into account the quantum-mechanical implications. The electrons of the electron gas show a totally irregular movement due to the thermal energy. Thus the different movements of single electrons compensate each other statistically and no macroscopic electric current is observed. This changes when an electric field is applied. Due to the electric field occurs a drift of the electrons along the field and one observes a current. The electrons collide with the atoms of the crystal lattice. Between two collisions they are accelerated in the electric field, they get additional kinetic energy,

30 D.M. Kolb, J. Schneider, Electrochimica Acta 1986, 31, 929.

which they may lose in the following collision. The time between two collisions can be defined by

$$t_s = l/v_s \tag{2.202}$$

where t_s is the averaged time between two collisions and v_s is the averaged thermal velocity of the electrons. The movement of the electrons in the electrical field finally leads to the macroscopic, observable current. This current follows the Ohmic law $i = U/R$ or $i = \kappa\, U$.

The energy band model

This model is the standard model of solid state physics. In the following only a short introduction to electrons in solid will be given. A more detailed representation of this topic is given in Appendix D.

In solids, no discrete atoms or molecules exist. A bond network extends throughout the volume of a single crystal. As known from molecular orbital theory two degenerate energy levels which exist between a pair of neighboring atoms split for a discrete bond in a bonding and an antibonding level. The bonding level is occupied by two electrons, and the antibonding level is empty. In a macroscopic solid the atomic levels split into a very large number of closely spaced energy levels (Fig. 2.41). This ensemble of energy levels is called an energy band.

In an energy band, the energy spacing between adjacent energy levels is so small that energy bands are treated as a continuum of energy levels. This continuum has well-defined upper and lower energetic limits, the band edges. The

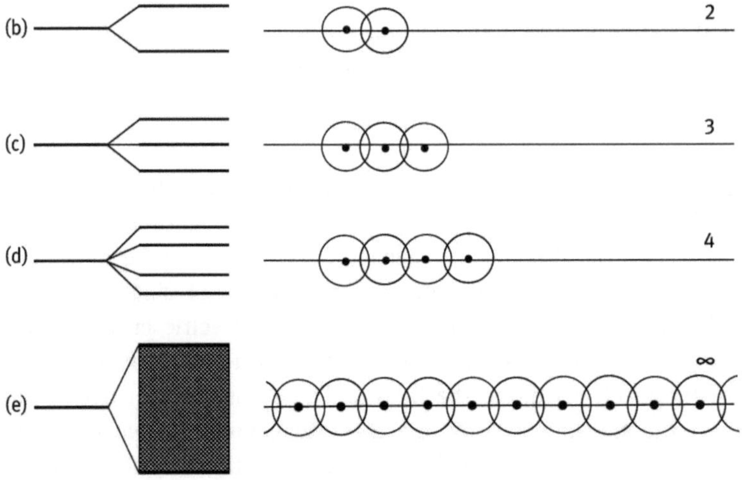

Fig. 2.41: Formation of electron bands of a solid by splitting of electron terms of interacting isolated atoms.

highest with electrons filled band is called the valence band, the lowest empty band is called conduction band. The highest filled energy band and the lowest empty energy band are the most important in terms of charge transport and charge transfer. In a semiconductor or an insulator the bands are separated by a gap devoid of energy levels. In a metal the two bands overlap or a band is partially filled with electrons.

Fig. 2.42 shows some possible situations arising in the frame of the band model. The band structure of a metal with half-filled valence band and an empty conduction band is shown in Figure 2.42a. The second situation (Fig. 2.42b) is characterized by a very large gap between valence band and conduction band, so that no electrons can be transferred into the conduction band. This is typical for insulators. Figure 2.42c shows a band structure with completely filled valence band and an empty conductivity band. The distance between these bands is small. Since the valence band is filled and has no empty energy levels in which electrons can move, only the excitation of electrons into the conduction band delivers mobile electrons. At room temperature thermal energy is sufficient to transfer electrons from the valence band into the conduction band and to create thereby electric conductance. This situation is typical for intrinsic semiconductors, like silicon or germanium.

In most common semiconductors the valence band is derived from a bonding molecular orbital and the conduction band from an antibonding molecular orbital.

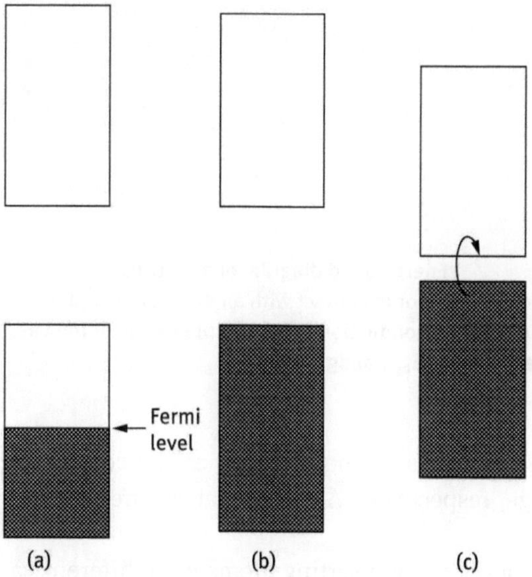

Fig. 2.42: Comparison of the band model for a metal (a), an intrinsic semiconductor (c) and an isolator (b).

This has important consequences when charge carriers reach the semiconductor/electrolyte interface.

Charge carrier generation

A metal conducts electricity because it has empty energy levels immediately above the occupied ones. Electrons are thermally excited to the empty levels and acquire an electric field mobility along the adjacent empty levels.

A perfect intrinsic semiconductor at 0 °K is a nonconductor because the levels of the valence band are completely occupied and those of the conduction band are totally empty. To impart electrical conduction, charge carriers must be created.

There are three mechanisms for creating charge carriers: thermal generation, photo-excitation and doping.

Provided the band gap is small, by thermal excitation electrons from the valence band are transferred into the conduction band, as shown in Fig. 2.42c. Consequently, a positive charge is created in the valence bands. Both the electron and the positive charge are mobile within their respective energy bands. The concentration of electrons (n) and holes (p) depends on the absolute temperature and the band gap energy E_{bg} (Fig. 2.43):

$$(n)(p) = N_c N_v \exp(-E_{bg}/kT)$$

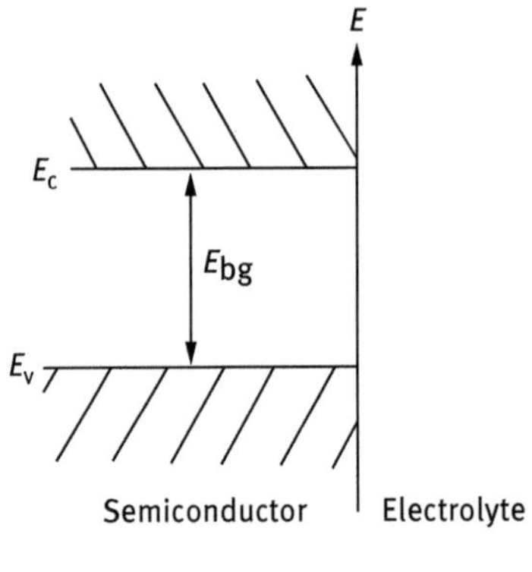

Fig. 2.43: Energy band diagram of an intrinsic semiconductor in contact with an electrolyte. E_c lower edge of the conduction band, E_v upper edge of the valence band, E_{bg} bandgap energy.

where N_c and N_p are the “effective” densities of energy levels at the conduction band edge and the valence band edge, respectively. At room temperature their values are of the order of 10^{19} cm^{-3}.

In doping, the crystal lattice is modified by inserting atoms with different valencies. In a crystal composed of atoms with four valencies, for example Si, electron donating atoms like P or electron deficient atoms like B can be inserted. This leads to *n*- or *p*-semiconductors, respectively. In this way, energy levels are introduced in the band gap region. More details can be found in the Appendix.

The Fermi level

The Fermi level is an extremely important parameter for metals and semiconductors. It is the property which controls the externally applied potential. The Fermi level is the electrochemical potential of the electron in the solid. Alternatively it is defined: The Fermi level is the energy E_f at which the property of an energy level being occupied by an electron is ½.

The interface semiconductor electrolyte

At the interface between a semiconductor and an electrolyte, the double layer consisting of the Helmholtz layer and the diffuse layer exist at the side of the electrolyte as in metals.

However, in the semiconductor side develops a space charge layer as soon as an external potential is applied. This layer contains excess electric charge and this leads – following the Poisson equation – to a modification of the band energies. At the surface the bulk energy band of the conduction band and of the valence band is bend upward or downward, depending on the nature of the space charge near the surface. This is shown in Fig. 2.44.

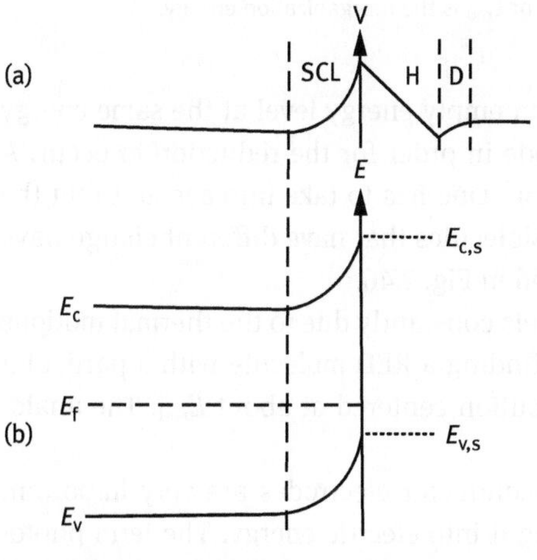

Fig. 2.44: Drops of the electrical potential across the semiconductor interface and band-bending. The upper curve (a) shows the electrical potential across the interface: SCL, space charge layer near the semiconductor surface; H, Helmholtz layer; D, diffuse layer. The relative thickness of these layers is not shownin (b) E_{vs} and E_{cs} mark the surface energy of the band edges of the valence band and the conduction band. E_f is the energy of the Fermi level. The semiconductor is intrinsic (undoped) with positive band bending.

Electron transfer – a new model

At a metal electrode, electrons are transferred between the depolarizer in the electrolyte solution and the electrode within a single energy band. In contrast, in a semiconductor the electron transfer can take place via two energy bands.

This fundamental difference provides an opportunity to develop and test the essential ideas of electron transfer mechanism. A unifying model, the "fluctuating energy level model", exists for predicting electron transfer kinetics at an electrode. Its application to semiconductors has been pioneered by Heinz Gerischer.[31]

The model is based on the assumption that the electron transfer is too rapid for any significant molecular motion to occur. The transfer of an electron between an electrode and a molecule in the electrolyte takes place before the (solvated) molecule can adjust to its new charge. The electron moves horizontally across the interface in an energy-level diagram (Fig. 2.45).

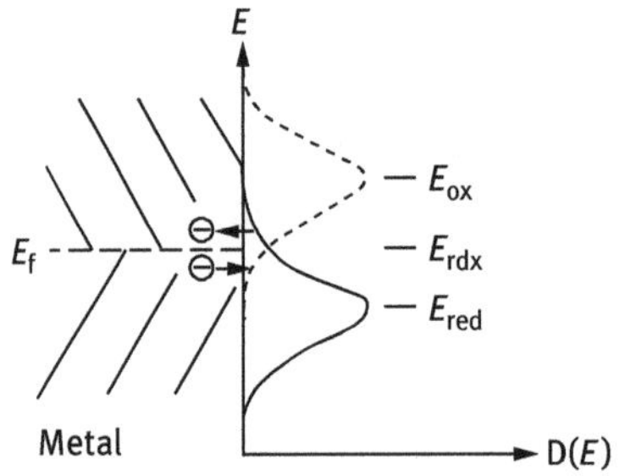

Fig. 2.45: The fluctuating energy level model at a metal electrode. $D(E)$ density of state of redox energy levels,; [Ox] = [Red]; $V = V_{rdx}$ electrode at the Nernst potential; E_{red} maximum density of occupied redox levels, E_{Ox} maximum density of empty redox levels. The difference between E_{rdx} and E_{Ox} or E_{red} is the reorganization energy.

For reduction the molecule must possess an empty energy level at the same energy as an occupied energy level in the electrode in order for the reduction to occur. A similar statement can be made for oxidation. One has to take into account that the molecules in the electrolyte are solvated. Molecules that have different charge have different solvation energy. This is illustrated in Fig. 2.46.

The molecules change their energy levels constantly due to the thermal motions of the solvent dipoles. The probability of finding a RED molecule with a particular energy level is given by a Gaussian distribution centered at about E_{red}. The analogous situation holds for OX molecules.

Electrochemical cells containing semiconductor electrodes are very important for harvesting light energy and transferring it into electric energy. The term photoelectrochemistry meets this part of electrochemistry. The so-called Grätzel cell will be described later.

31 H. Gerischer in "Advances in Electrochemistry and Electrochemical Engineering" Vol. 1, P. Delahay (ed.), Interscience Publishers, New York 1961, pp. 139–232.

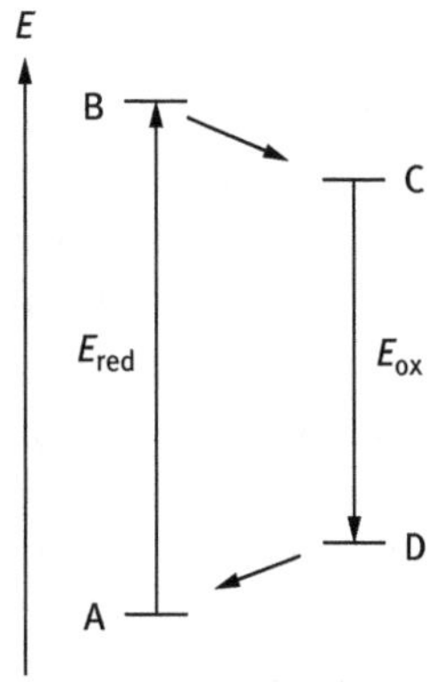

Fig. 2.46: Thermodynamic cycle for determining the magnitudes of E_{red} and E_{ox}. State A: The energy of a RED molecule with a Red solvation structure. State B: The energy of an Ox molecule with a Red solvation structure. (An electron has been removed to infinity.) State C: The energy of an Ox molecule with an OX solvation structure; State D: The energy of a Red molecule with an Ox solvation structure (The electron has been returned from infinity) $E_{red} = E_b - E_a$; $E_{ox} = E_d - E_c$.

Photoelectrochemistry

One speaks of photoelectrochemistry if a current is induced in an electrode/electrolyte system excited by light. The photocurrent is observed between the light-sensitive electrode and a counterelectrode. Photocurrents are observed preferably on semiconductor electrodes. The amount of photocurrent on a given electrode depends on the potential of the electrode, the composition of the electrolyte and last not least on the energy of the incoming radiation, i.e., the wavelength of the light.

The photocurrent is generated by the excitation of electrons in the valence band into the conduction band. Figure 2.47 illustrates the situation at an *n*-doped semiconductor.

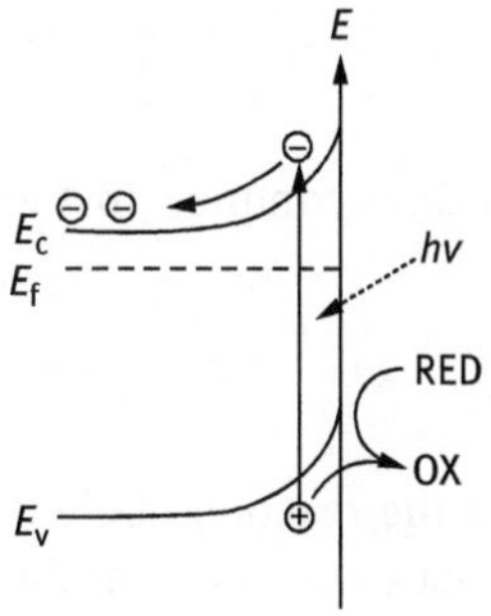

Fig. 2.47: Photocurrent generation at an *n*-type semiconductor. E_f, energy of the Fermi level; the photogenerated positive holes move to the surface and oxidize solution molecules.

The dye-sensibilized solar cell, often called “Grätzel cell,” is a special and prominent example for the application of photoelectrochemistry. It will now be described.

The cell contains three main components:

The photo electrode, where light absorption occurs and the charge carriers are generated.

The counter electrode where the oxidized species dissolved in the electrolyte are reduced.

The electrolyte that is responsible for the charge transport between the two electrodes.

The Grätzel cell uses titanium dioxide as a semiconductor. This material is cheap and easily available. Another factor that contributed to the popularity of this material was the ease of fabrication polycrystalline electrodes. A prerequisite for the efficiency of the cell is very efficient absorption of light. This is not possible with neat TiO_{2}., because the energy gap between the valence band and conduction band is 3.05 eV for rutil, and the onset wavelength for photocurrent generation is 410 nm. The band gap for anatase is 3.20 eV. These data are not optimal for light harvesting.

In the Grätzel cell, the semiconductor electrode (*n*-doped anatase) consists of a thin layer of nanocrystalline TiO_2 particles (diameter 10–25 nm) on optically transparent FTO glass (fluorine-doped tin oxide). On the TiO_2 layer a specific dye is adsorbed with a broad absorption spectrum in the range of the visible incident solar radiation. Ruthenium complexes are the best choice for this purpose.

The counterelectrode is FTO glass coated with a thin layer of platinum. The electrolyte contains the complex I_2I^-.

Four fundamental processes that occur in the cell are

$$S + h \cdot \nu \longrightarrow S^*$$

The dye (sensitizer S) absorbs light and is transferred to the excited state S*:

$$S^* \longrightarrow S^+ + e^- (TiO_2)$$

Injection of the electron into the semiconductor yields

$$I_3^- + 2e^- \longrightarrow 3I^-$$

After passing the outer circuit the electron reduces the iodine complex in the electrolyte:

$$S^+ + I^- \longrightarrow S + 1/2\ I_2$$

Finally, the oxidized dye is reduced by the iodine anion and the reaction circle is ended. In the cell, these processes occur simultaneously and not sequentially as described earlier.

Due to loss mechanisms the efficiency of the cell is limited. The main loss reactions are the following:

$$S^* \longrightarrow S + h \cdot \nu$$

The excited sensitizer returns to its ground state before the electron has been injected into the semiconductor:

$$e^-(TiO_2) + S^+ \longrightarrow S$$
$$2e^-(TiO_2) + I_3^- \longrightarrow 3I^-$$

The injected electron may be transferred to the dye cation S^+ or directly to I_3^- in the electrolyte.

Under optimal conditions the open-circuit voltage of the cell reaches 0.7–0.9 V and the efficiency near 11 %. These values are near those obtained in the silicon technology.

Charge transfer by ions in solids

In ionic crystals, like NaCl, the electrical charge is localized at the ions. No mobile electrons exist. Nevertheless, transport of electrical charge can occur if there are vacancies in the crystal lattice. By hopping of ions into a neighboring vacancy and creating a new vacancy the ions can move through the crystal following an electric field gradient. Since hopping of a cation creates a negatively charged vacancy, this vacancy migrates in the direction opposite to that of the cation migration.

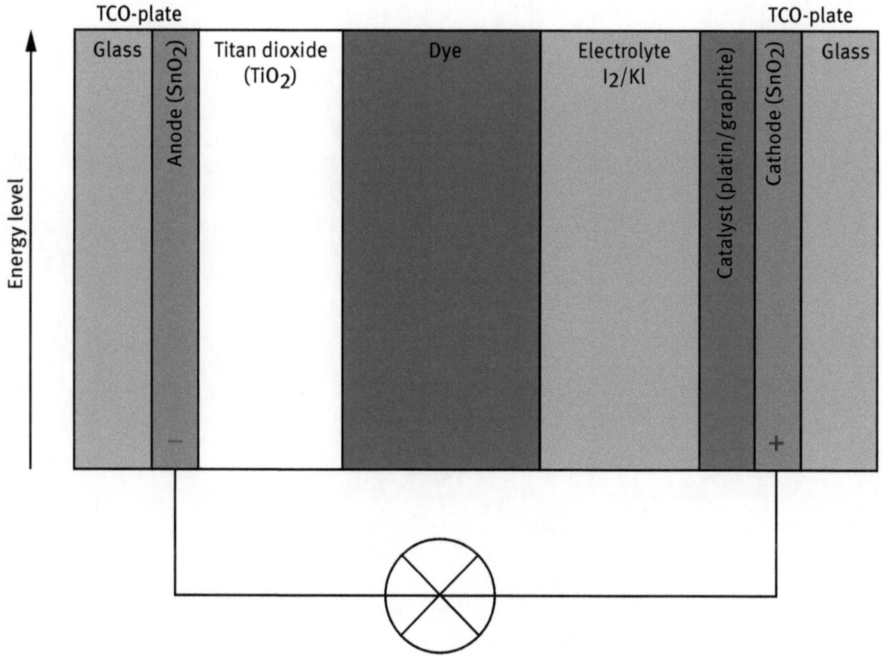

Fig. 2.48: Schematically shows the Grätzel cell (Source: Jonas Gabriel).

Conclusion

At this point we arrive at the end of the chapter. I hope the reader has now a feeling for the different elementary processes involved in electrode reactions.

We have described only the very basic systems using simple macroscopic models. Many interesting and important systems for applications, for example, solid state electrochemistry and semiconductor electrodes have been omitted. They are out of the scope of this book, which shall be an introduction guiding newcomers to understand the basic processes of electrochemistry.

The next part of this book is dedicated to the instrumentation of electrochemical experiments and to some experimental methods often used by electrochemists.

3 Practical work in electrochemistry

3.1 Instrumentation and methods

In this chapter, we discuss the practical use of the fundamentals of electrochemistry described in the preceding chapters.

For the investigation of an electrochemical system as well as in applications of electrochemistry exist a unique principle: We send a controlled input into the system and measure the response of that system.

The controlled input may be the potential at the working electrode and the response may be a Faraday current. The resulting current-potential (I–E) curve contains the relevant information. The I–E curves may have very different shapes depending on the nature of the electrode process and the form of the input signal. The interpretation of I–E curves will be the main subject of the following chapters. Using a rotating disk electrode (RDE), in addition to the potential, we may control the mass flow to the electrode.

In a potentiometric titration, a very old analytical technique, the composition and concentration of a solution is controlled. The response is either the potential of the electrode or the conductivity of the solution. The general concept is as follows:

$$\textbf{Excitation} \rightarrow \textbf{System} \rightarrow \textbf{Response}$$

Box 5
Parameters influencing the properties of the interface electrode/electrolyte

Material of the electrode

- Different metals and semiconductors

Surface parameters

- Pure surfaces of single crystals, surface structure, reconstruction
- Polycrystaline surfaces,
- Liquid surfaces
- Structure of the surface, roughness, true area
- Covered surfaces, physisorption, chemisorption
- Thickness and structure of adlayers
- Charge density at the surface
- Concentration of electroactive species at the surface

Solution parameters

- Solvent: water, organic liquids, liquid electrolyte, etc

https://doi.org/10.1515/9783110437393-004

- Concentration of electroactiv species (c_O, c_R)
- Concentration of surface active species
- Mass transfer (diffusion, convection)

Experimental parameters

- Electrical parameter (potential, current, transfered electrical charge)
- External parameter (Temperature, pressure, time)

A short compilation of variables influencing the electrode (see Box 5) processes will be provided.

The electrical variables are potential (E), current (i) and the quantity of electricity (Q). The electrode itself can vary depending on its material, surface and geometry; these variables do not affect the potential of the electrode. Further variables are connected with the solution. These are the concentrations of electroactive species (c_O, c_R), the concentration of other species, such as the supporting electrolyte, p_H and finally the solvent. The variables that are concerned with the mass transport are diffusion, migration, convection, adsorption and surface concentration. Following are the external variables: temperature, pressure and, last not least, time.

The approach of the experimenter should be to apply a suitable excitation function (e.g., a potential step) and to measure a certain response function (e.g., the decay of the current in time) with all other system variables being constant. The aim of every experiment is to get thermodynamic, kinetic, analytical information, and so on about the system. Finally an appropriate model of the system may be designed.

3.2 The ingredients of an electrochemical experiment

An electrochemical experiment involves electrodes, an electrolytic solution and a suitable vessel. In addition to these, a source for variable potentials (electrostat plus potential sweep generator) and an amperemeter and a voltmeter with high internal resistance complete the experimental equipment. Three electrodes are used in most of the experiments.

3.2.1 Cell design

For practical use there exists a manifold of different cell designs. A standard cell should contain five ports, three for the electrodes and two for the entrance and exit of an oxygen-free inert gas such as N_2 or Ar, bubbling through the electrolyte and deoxygenating the solution.

Some different equipment for different kinds of electrochemical experiments are shown in Fig. 3.1.

A standard vessel has at least three standard ground joints for the working electrode, the reference electrode and the counter electrode. Two additional ground joints connect to the gas inlet and outlet valves. The solution under investigation has to be deaerated, because the reduction of oxygen often disturbs the experiment. This can be done by bubbling a purified inert gas (argon, nitrogen) through the solution.

The cells are usually constructed in glass; special materials are used only for working with highly corrosive substances such as liquid HF or with molten salt electrolytes. The reference and the counter electrodes have separate compartments that are separated from the solution under investigation by permeable sintered-glass disks or other usable materials. The distance between the working and the

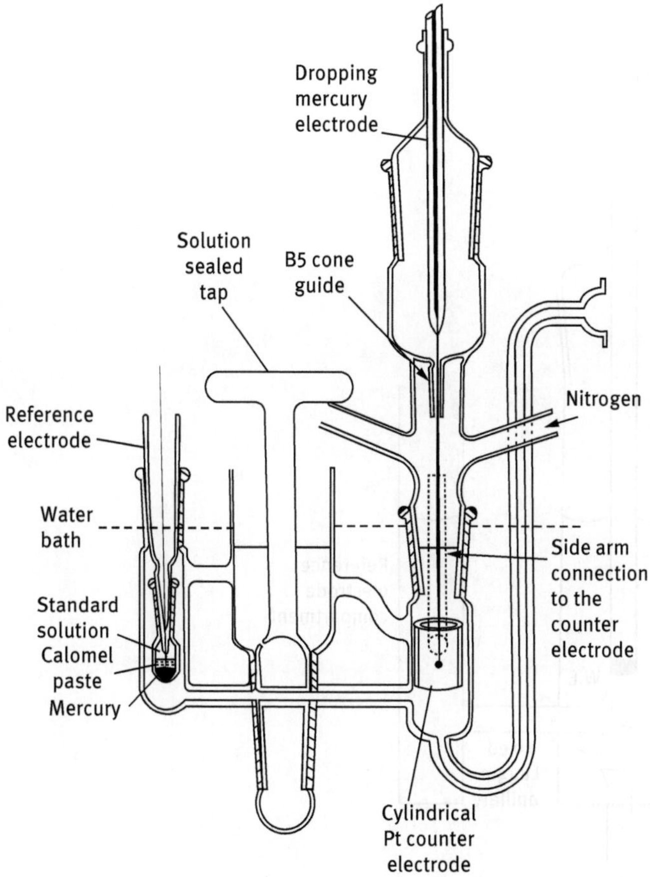

Fig. 3.1: Some devices for performing different electrochemical experiments.

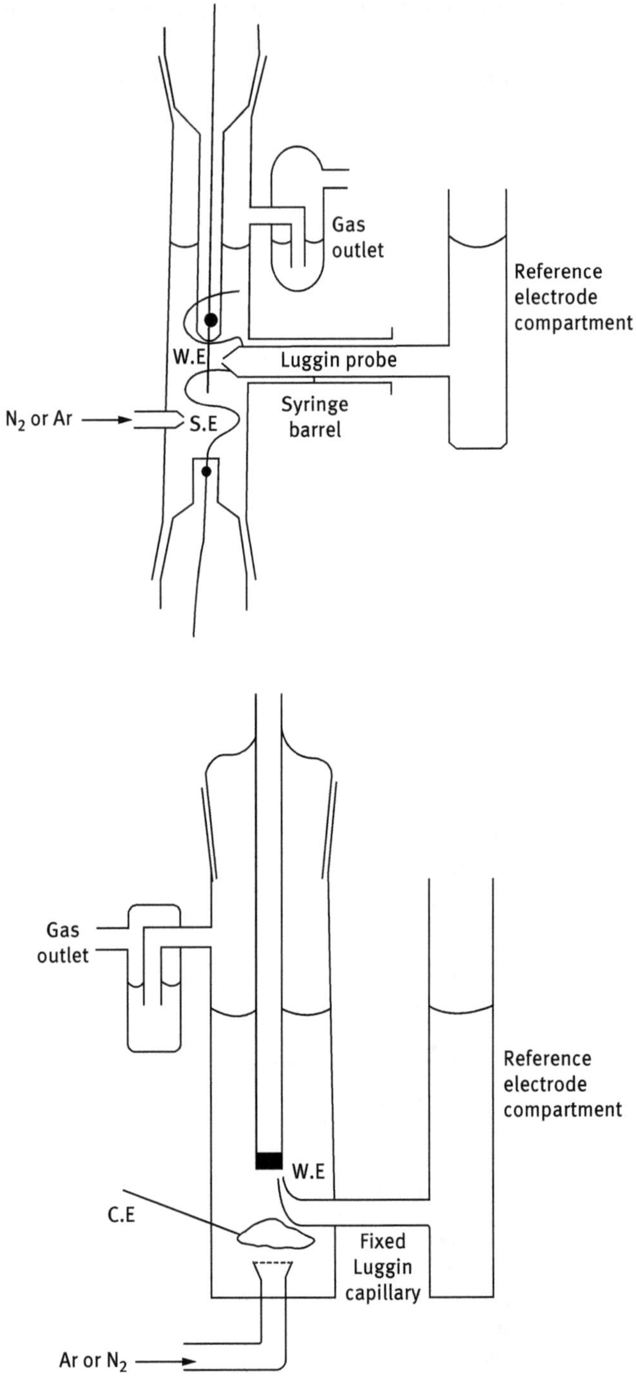

Fig. 3.1 (continued).

counter electrodes should be small. This minimizes the resistance for the current driven through the cell by the potentiostat.

However, optimizing the geometrical configuration of the electrodes is not sufficient for establishing the required potential at the working electrode. This potential is influenced by the current flowing between working and counter electrodes and the conductivity of the solution. The actual potential V is in error by an amount of:

$$V_{iR} = i \cdot R_u \tag{3.1}$$

where i is the cell current and R_u is the uncompensated resistance between the electrode and the probe tip (Fig. 3.2).

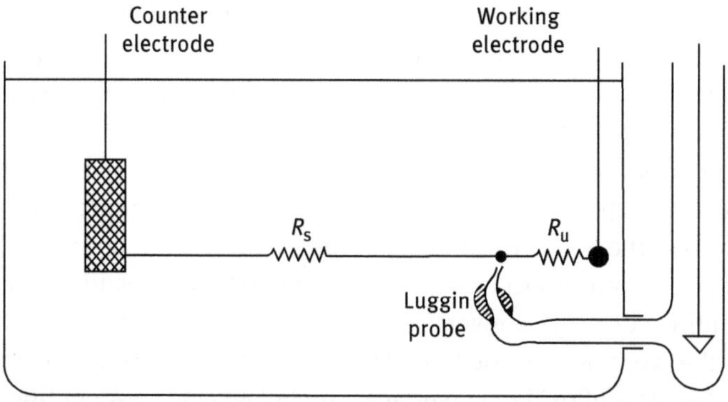

Fig. 3.2: Schematic diagram of a cell showing the principle of the Luggin probe. R_s is the resistance of the solution between the counter electrode and the tip of the probe. R_u is the uncompensated resistance between the Luggin tip and the working electrode surface.

The Luggin capillary is the device for minimizing errors in the potential because of the uncompensated resistance of the solution. The diameter of the Luggin capillary and the position of its tip that are close to the working electrode; it must be optimized experimentally. Useful tips for doing this can be found in the literature.[32]

The working electrode defines the interface under study, the reference electrode maintains a constant reference potential. The counter electrode is an auxiliary electrode that supplies the current required by the working electrode without limiting the measured response of the cell. When the working electrode is an anode, the counter electrode is a cathode and vice versa. Generally the electrode process at the counter electrode is of minor interest. The electrodes are separated in space by a glass frit, a membrane, or something else. In the following sections the cell components are described in detail.

32 Instrumentation Methods in Electrochemistry, Ellis Horwood Limited 1985.

3.2.2 The working electrode

The working electrode may be a small disk, a short wire, or a sphere (mercury). For the realization of the electrode that exists has principally no limits; however, in practice, small electrodes with an area of about 0.3 cm^2 are used. A special case is that of ultramicroelectrodes that are not considered here. The materials of the electrode will not react chemically with the solvent or the solution components. For solid electrodes, gold and platinum are preferred materials; the dropping mercury electrode has a reproducible clean and smooth surface, but it can be hardly used for investigation of anodic reactions. Other materials preferred are glassy carbon and lead. Semiconductors and evaporated films are also used for working electrodes. For in situ transmission and photocurrent spectroscopy, thin glass slides coated with transparent, conducting SnO_2 are available. During the past two decades, single crystal electrodes were of special interest because they allowed the detailed study of electrochemical reactions on a molecular level. Single crystal gold electrodes are nowadays easily available because of a procedure first published by Hamalin.[33]

Even if polycrystalline working electrodes are used, the surface should be smooth because the mass transport and the geometry are then better defined. Generally, the electroactive and the geometrical areas of the electrodes are not identical. The electroactive area of an electrode can be determined by measuring the diffusion controlled limiting current of a reversible electrode process.

A pretreatment of the working electrode before using it is often necessary. In order to get a reproducible state of oxidation, surface morphology and freedom from adsorbed species the solid electrodes should be polished. Polishing is performed on wet cloth pads impregnated with diamond particles down to 1 μm grain size and followed by alumina of grain size down to 0.05 μm for easy recognition. It has been observed that diamond grains sometimes contain dyes that may be electroactive. Small particles of alumina may stick in the soft metal surface and influence the measurement because of the adsorption of the reactant at the surface of the alumina. After polishing, the electrode should be kept into an ultrasonic bath for a short period of time. For producing a smooth surface, chemical etching is often used. The etches are usually strong oxidizing solutions (e.g., mixtures of nitric and sulfuric acids) producing a high oxidation state of the material that dissolves in the aqueous phase.

Finally, the electrodes should be polished electrochemically, for example, by cycling the potential between hydrogen and oxygen evolution, thereby surface impurities may be oxidized.

A convenient way of mounting solid electrodes is to melt a rod of the material into a glass tube such that only the circular cross section is exposed. The sealing of the glass can also be done by using epoxy resin or plastics such as teflon or "Kel-F."

---

33 A.Hamalin in "Modern Aspects of Electrochemistry" Plenum Press, New York, 1985, Vol 16, p. 1ff.

3.2.3 The reference electrode

Reference electrodes are nonpolarizable electrodes that provide a fixed potential. The potential of the reference electrode must be stable during the experiment. Ideal nonpolarizable electrodes are hardly available. A good reference electrode should maintain a constant potential even if a few microamperes pass through its surface. As an example, if $I < 10\ \mu A$ and $R < 100\ \Omega$ we get $iR < (10^{-5}\ A)(R < 100\ \Omega)$ or $iR < 1mV$.

If the potential between working and reference electrodes is controlled by a potentiostat (see below) then any change in the applied potential appears directly across the interface of the working electrode/solution. Being related to the normal hydrogen electrode, the agreed standard for thermodynamic calculations and the reference electrode provide a thermodynamic standard. As mentioned previously the reference electrode is isolated from the working electrode by a glass frit.

In the following sections, we give a short compilation of the commonly used reference electrodes. The theory and characteristics of reference electrodes have been reviewed by Janz and Ives.[34]

The calomel electrode (mercury/mercurous chloride) is a widely used reference electrode. It is usually surrounded by a saturated solution of potassium chloride. The abbreviation SCE stands for Saturated Calomel Electrode. Sometimes a 1 molar or 0.1 molar aqueous solution of KCl is used. Similar reference electrodes can be prepared with mercurous sulphate. For use in alkaline solutions the mercury/mercury oxide electrode is recommended.

The silver/silver chloride electrode is also widely used. It can be prepared easily by electrochemical coating a silver wire with silver chloride and dipping it in a KCl solution. The Ag/Ag^+ potential is fixed by the solubility product $L = a\ (Ag^+)(Cl^-)$ of silver chloride. This reference electrode can also be used in many nonaqueous electrolytes.

The potential of reference electrodes changes when nonaqueous electrolytes are used.

The relation between the potentials of reference electrodes in different solvents can be obtained by using the redox potentials of ferrocene/ferrocenium or cobaltocene/cobalticinium as a reference potential. In these compounds the central metal atom is surrounded by big organic ligands and in this way it is shielded against the solvent molecules. The redox potentials of these couples measured on a platinum electrode are independent from the solvent. Thus measuring the potential of reference electrodes in different nonaqueous electrolytes versus the redox couple cobaltocene/cobalticinium, one finds the shift of the reference potential because of the change in the solvent of the electrolyte. In Table 3.1 the potentials of some reference electrodes are compiled.

34 G.J. Janz, D.J.G. Ives, "Reference electrodes", Academic Press, 1961.

Table 3.1: Potential of some reference electrodes in aqueous solutions at 298 K vs. NHE**. It contains secondary and redox electrodes.

Name	Electrode	Potential versus NHE**
SCE*	Hg/Hg_2Cl_2, sat. KCl	+0.241
Calomel	Hg/Hg_2Cl_2, 1 molar KCl	+0.280
Mercurous sulphate	Hg/Hg_2SO_4, sat. K_2SO_4	+0.640
Mercurous oxide	Hg/HgO, 1 molar NaOH	+0.098
Silver chloride	Ag/AgCl/, sat. KCl	+0.197
Silver bromide	Ag/AgBr/, sat. KBr	+0.071
Silver iodide	Ag/AgJ/, sat. KCJ	−0.152

* Saturated Calomel Electrode.
** Normal Hydrogen Electrode.

3.2.4 The counter electrode

For most of the electrochemical experiments, "three-electrode" setups are used. The reason is that in a solution where *iR* may be high, a current flowing in a "two-electrode cell" induces an overpotential on the working electrode. This makes it necessary to use an auxilliary electrode for precise measurements, which is the counter electrode. In such an arrangement the current is passing between working and counter electrodes. This avoids an additional overpotential because of *iR*; the counter electrode should be large in comparison with the working electrode. It is separated from the electrolyte by a sintered-glass disk. The material of the counter electrode may be gold, platinum, or any other desired material, because substances produced in electrolysis at the counter electrode do not affect the area of the working electrode.

3.2.5 The electrolyte solution

In electrochemical experiments the solution generally contains the supporting electrolyte at a rather high concentration (ca. 0.1 mol/L). This is a well-dissociated salt, which is not electroactive in the potential range under investigation. The supporting electrolyte increases the conductivity of the solution and hence reduces the resistance between working and counter electrodes. It reduces the power requirement on the potentiostat and minimizes the potential error arising from the uncompensated resistance iR_u. The concentration of the supporting electrolyte exceeds strongly that of the electroactive species ($\approx 10^{-3}$ mol/L) and eliminates the migration as mass transport of the electroactive species. This facilitates the understanding of the current time relation in cyclic voltammetry (CV) and other potential sweep techniques. The salts used as supporting electrolytes have to be purified carefully by several recrystallizations. If water has to be excluded from the experiment, then purified supporting electrolytes should be stored under an inert gas atmosphere

The choice of an appropriate solvent for an electrochemical system is as meaningful as the choice of the supporting electrolyte. Similar to the supporting electrolyte the solvent may limit the potential range of a planned experiment. Of course, water is mostly used as a solvent. Salts are properly dissolved in water; the dielectric constant is rather high (DK = 80). This supports the dissociation of salts. The dielectric constant is often used as a first preference for the dissociation of salts; however, one should bear in mind that it is the interaction with solvents that governs the dissociation of salts and the stability of solvated ions.

Impurities in the solution, even at very low concentrations ($<10^{-4}$ mol/L) may disturb measurements considerably, often by adsorption at the working electrode. Therefore careful cleaning of the solvent is necessary. Water has to be distilled several times prior to use (from dilute potassium permanganate, phosphoric acid, etc.). It must be deoxygenated by bubbling nitrogen or argon through the solvent. It is advantageous to perform all the experiments under a protective gas atmosphere (N_2, Ar).

During the past decades the electrochemistry of organic systems became the subject of increasing interest. This requires organic solvents. Not only the solubility of organic compounds but also the usable potential range recommends the use of organic solvents.

These solvents also need to be purified carefully, mainly by an appropriate chemical pretreatment and fractional distilllation. Acetonitrile became the leading solvent in organic electrochemistry because of the DK of 36 and the wide "open" potential range. However, some experience is needed to produce and handle non-aqueous acetonitrile. Alternatively, benzonitrile (DK 25), dimethyl sulfoxide and dimethylformamide may be used.

In Table 3.2 some solvents for electrochemical experiments are compiled.

In aqueous medium the salts of alkali ions are generally used as supporting electrolytes; in organic solutions alkylammonium ions NR_4^+ (R = CH_3, C_2H_5, C_3H_7, n–C_4H_9, etc.), X^- (X = F^-, Cl^-, ClO_4^-, BF_4^-, etc.) are the favored supporting electrolytes.

3.3 Instrumentation

In electrochemistry mostly electronic instruments are used for controlling and measuring the charge (coulostat, integrator), current (galvanostat, current follower)

Table 3.2: Solvents and supporting electrolytes for electrochemical experiments.

Solvent	Potential range[1]	Supporting electrolyte	DK[2]	Application
Water	Hg: −1.9 to 0, Pt: −0.7 to 1.1	Different salts NaF, $LiClO_4$, etc.	78.3	Aqueous solutions of salts,various pH, metal complexes
Ethanol, methanol	–	Different salts, for example, $NR_4^+X^-$ R = alkyl and X= F^- ClO_4^- BF_4^-	24.3	–
Acetonitrile	Pt: −1.5 to 1.6 Hg: −2.8 to 0.6	NR_4X R = Alkyl, X = F^-, ClO_4^-, BF_4^-, etc.	36.3	Nonaqueous solutions
Benzonitrile	–	NR_4X R = Alkyl, X = F^-, ClO_4^-, BF_4^-, etc.	26	Aprotic solutions
Dimethylsulfoxide	Pt: −1.9 to 0.75	For example, $NaClO_4$	46.6	Aprotic solutions
Dimethylformamide	Pt: −1.6 to 1.6	For example, $NaClO_4$	36.5	Aprotic solutions
Propylene carbonate	Pt: −1.85 to 0.70 Hg: −2.85 to −0.40	$(C_2H_5)_4CLO_4$	64	Organic solvents
Dioxane	–	$(C_4H_9)_4CLO_4$	2.2	–

[1] V vs. NHE
[2] Dielectricity constant

and the potential (potentiostat, voltage sweep generator and high impedance voltmeter). These instruments may be linked to a digital computer, which controls the course of the experiment.[35]

3.3.1 Why are three electrodes used in an experiment?

In an electrochemical cell at least two electrodes (working electrode, either anode or cathode and a reference electrode) are needed to perform current flux through the cell. In a two-electrode system the current and the cell voltage between anode and cathode can be measured and adjusted. However, it is not possible for the potential at the working electrode alone. However, the measurement of this potential is important because it governs the electrode reaction. For the measurement of the potential of the working electrode, a third electrode is necessary; this is the nonpolarizable, current-free reference electrode.

35 Many details in chapters 3.3 and 3.4 have been taken from “Instrumental Methods in Electrochemistry”, Southampton Electrochemistry Group, Ellis Horwood Limited, Publishers Chichester, 1985, p. 370ff.

3.3.2 Potentiostat

Among the different devices used in electrochemistry the potentiostat is undoubtedly the most important one. Only the basic principle of a potentiostat will be reported.

To realize at the working electrode *M*, a required potential $\varphi_M(i)$ needs the continous comparison with a reference potential; this is generally the potential of the reference electrode B or the potential at the end of the luggin capillary.

With the Luggin capillary (potential probe) the potential is registered very close to the working electrode. This signal is put into the potentiostat (Fig. 3.3). and here the measured potential U_{ist} is compared with the required potential U_{soll}. The potential U_{soll} may be a constant value or a value variable in time and generated by a voltage sweep generator. If a difference exists between U_{soll} and U_{ist} the control unit – an operational amplifier – aligns within the "rise time" of the potentiostat (about 10^{-5} s), U_{ist} to U_{soll} by sending a fitting current through the cell. This current is recorded at a resistance *R*. The corresponding potential $\varphi_M(i)$ by this method is used for adjusting the potential of the working electrode each time to the desired input potential U_{soll}.

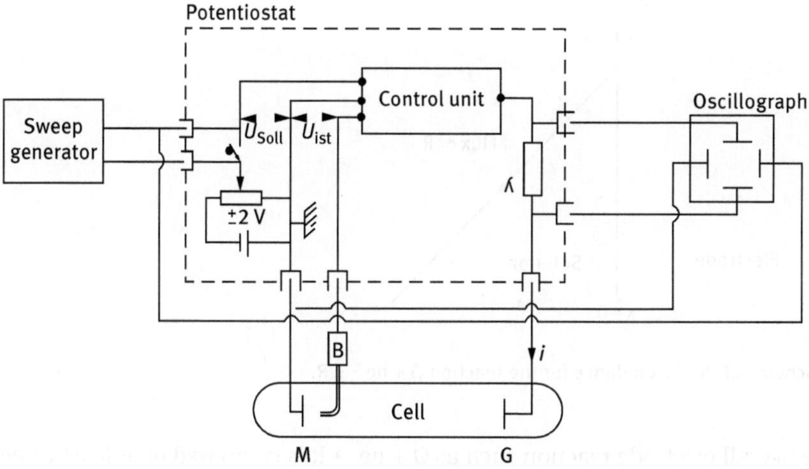

Fig. 3.3: Basic principle of a potentiostat.

If the input potential rises linearly in time (sweep generator) *I*–*E* curves can be measured directly and recorded by an oscillograph (or eventually by a fast X–Y recorder).

The first electronic potentiostat with rise times smaller than 10^{-1}s has been described by Gerischer and Vielstich.[36]

The galvanostat is a device that allows adjusting a constant current. Modern potentiostats combine both functions.

3.4 Experimental methods

There exists a wide variety of experimental methods for studying electrochemical reactions. Depending on the goal of an investigation, we can distinguish different categories of experimental procedures. For their description, we apply the fundamentals that have been described in the previous chapters. Therefore, it seems useful at the beginning of this new chapter to make some general comments on the fundamental constituents of electrochemical reactions.

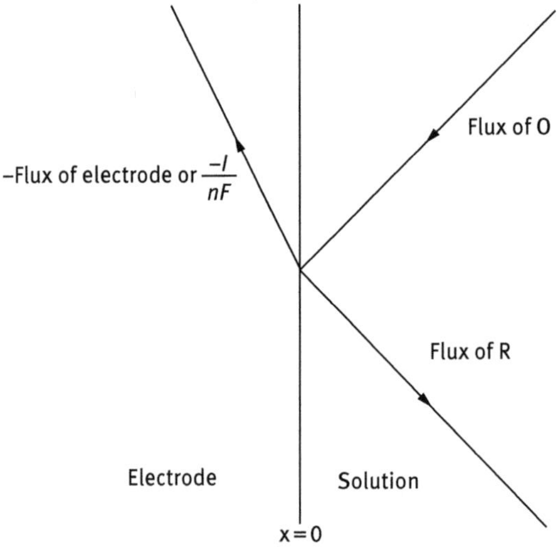

Fig. 3.4: Scheme of the flux balance for the reaction O + ne⁻→ R.

A simple overall electrode reaction such as O + ne → R is composed of at least three steps (fig. 3.4):

- Mass transfer of *O* from the bulk solution to the electrode surface
- Electron transfer at the electrode surface
- Mass transfer of *R* from the electrode surface into the bulk solution

36 W. Vielstich, H. Gerischer, Z. phys. Chem.N.F., 4,10 (1955).

The flux of O and R is given by diffusion. According to Fick's first law, the flux is proportional to the gradient of the concentration:

$$i/nF = -D_O(dc_O/dX)_{x=o} = D_R(dc_R/dX)_{x=o} \tag{3.2}$$

The observed current, that is, rate of the electrode process, is governed by the elementary steps.

In addition to these elementary steps, other processes may influence the appearance of the electrode reaction; these are as follows:

- Homogeneous chemical reactions such as protonation, dimerization, and so on preceding or following the electron transfer.
- Heterogeneous processes such as catalytic decomposition at the electrode surface.
- Other surface processes such as adsorption, desorption, or electrodeposition.
- Changes in the surface topology.

Complex electrode processes may involve a sequence of electron transfers and protonations, branching mechanisms, parallel paths or modifications in the electrode surface. The magnitude of the observed current is often limited by the inherent sluggishness of one or more reactions, called rate-determining steps. The evaluation of the mechanism of an electrode process can be very time-consuming.

Because of the great number of available methods, we are forced to select the considerations in the following text on some widely used methods. Here we focus on some of the methods that are used for analyzing the mechanism of electrode reactions. These are the measurement of current time curves at constant potential in an unstirred solution (chronoamperometry) or measurement at controlled mass transport using rotating electrodes. Special interest will be shown on cyclic voltammetry. In this method, current is the response to a time dependent voltage input.

The widely used methods such as coulometry, potentiometry and conductometry applied for the determination of concentrations of species will be discussed in the chapter on analytical "Applications of electrochemistry." Only some remarks are made on AC voltammetry.

As mentioned previously the principle behind performing electrochemical experiments is the controlling of a variable of the electrochemical cell and observing how other variables (usually current, potential and concentration) vary with changes in the controlled input variable.

The voltammetric experiment investigates the half-cell reactivity of the system by measuring the current at the working electrode as a function of the applied potential. The potential varies arbitrarily either continuously or step by step. The resulting curves $I = f(E)$ are called voltammograms. The shape of the voltammograms depends on the speed of potential variation and on whether the solution is stirred or quiescent.

3.4.1 Chronoamperometry

In this experiment, one uses a stationary planar electrode in an unstirred solution containing sufficient supporting electrolytes and the electroactive species. At the working electrode at $t = 0$ starts a fast potential jump to a constant potential E in the range of the limiting current of the electrode process. The response on the potential jump is a current–time curve as shown in Fig. 3.5.

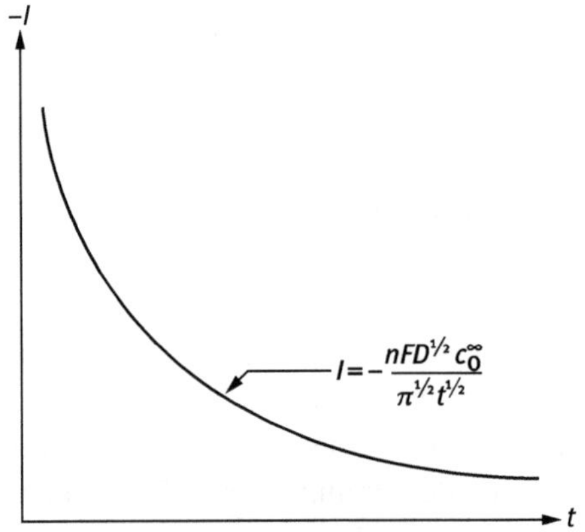

Fig. 3.5: Current–time response to the potential step described in Fig. 3.6.

The solution of Fick's second law (see appendix) leads to the exact form of the current–time curve (Fig. 3.5):

$$i = nFD^{1/2}\, c_O(\text{bulk})/\pi^{1/2}\, t^{1/2} \tag{3.3}$$

This curve can be used for evaluating diffusion coefficients and it confirms that the electrode reaction is diffusion controlled. Fig. 3.6 shows the corresponding concentration-time curves.

Measuring current–time curves, some experimental prerequisites have to be obeyed. The unstirred solution should contain a large excess of supporting electrolyte. Equation 3.3 is only valid on a time scale less than about 10 s and sufficient time must be left after each experiment for restoring the correct initial condition $c_O(0) = c_O(\text{bulk})$. At times smaller than 1 ms the current–time curve will be affected by the charging current of the double layer.

Equation 3.3 is a one-dimensional solution of Fick's second law; if a spherical electrode (mercury drop) is used the solution in spherical coordinates must be used.

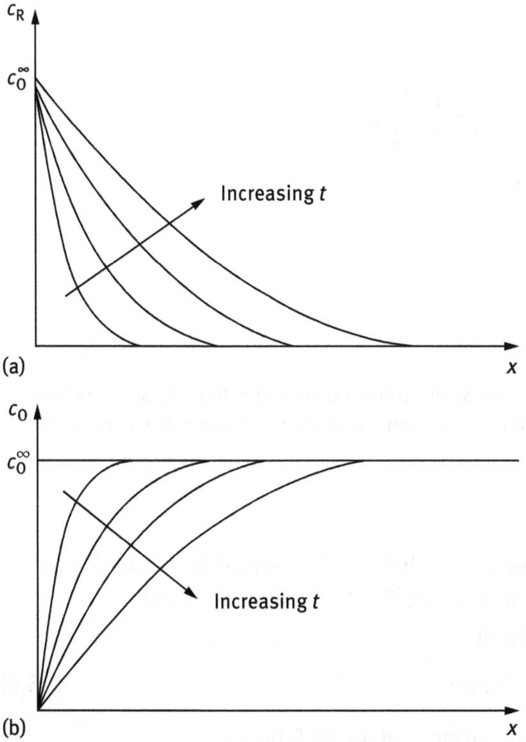

Fig. 3.6: Time evolution for the concentration profiles for the reaction $O + e^- \rightarrow R$ at a potential where the process is diffusion controlled, that is, for t>0, c_O = at x = 0; initially $c = c_O$(bulk) and $c_R = 0$ at all x.

3.4.2 Planar rotating disk electrodes

The RDE (Fig. 3.7) is realized by a small (diameter 1–2 mm) planar metal electrode, which is embedded in a suitable glass capillary. The distance between the electrode and the edge of the glass capillary should be at least ten times the radius of the electrode and the distance between the edge of the glass capillary and the wall of the cell is also to be taken into account.

As can be seen in Figures 3.7a,b the RDE acts as a pump. It sucks solution to the rotating plane and because of the centrifugal forces throws out the solution parallel to the plane. At the edge of the rotating disk the stream breaks off abruptly; this leads to turbulences in the solution.

With the RDE voltammetric measurements are performed in which the potential of the RDE is varied slowly in time, so that the equilibrium at the working electrode is established. The result of such a measurement is a characteristic I–E curve. The

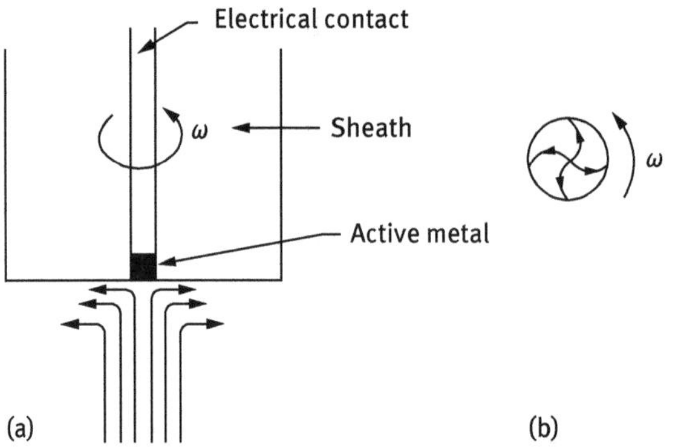

Fig. 3.7: Flow pattern created by the rotating disk electrode: (a) view from the side showing how the solution is pumped toward the disk and then thrown outwards, (b) solution flow close to the electrode surface viewed from below.

I–E curve can be described by an equation that can be derived by using the law of Nernst, and the relation for current at the RDE first published by Levich.[37]

We consider a simple redox system:

$$O + ne^- \rightarrow R \tag{3.4}$$

The equilibrium potential *E* at the electrode is given as follows:

$$E = E^0 + RT/nF \ \ln c_{O,O}/c_{O,R} \tag{3.5}$$

where E^0 is the redox normal potential, $c_{O,O}$ is the concentration of the oxidized form of the depolarizer at the surface of the electrode and $c_{O,R}$ is the concentration of the reduced form of the depolarizer. *R, T, n F* have their common meaning.

The current-free state of the rotating working electrode is described by eq. (3.5); it is valid for reversible electrode reactions in which the electron transfer across the interface of the electrode/electrolyte is very fast in comparison to the mass transport; in other words, the electrode process is diffusion controlled. If the potential at the electrode deviates from the equilibrium potential we observe a Faraday current *i*. Equation (3.5) is still valid if the electrode process is diffusion controlled.

Levich described the diffusion controlled cathodic current I_k with the following relation:

$$I_k = i_{l.k} - 0.62 \cdot n \cdot F \cdot A \cdot \nu^{1/6} \cdot \omega^{1/2} \cdot D_O{}^{2/3} \cdot c_{O,O} \tag{3.6}$$

and for the anodic current with the analogous relation

37 V. G. Levich “Physicochemical Hydrodynamics”, Prentice-Hall Inc., Eaglewood Cliffs, N.J., 1962, p. 60.

$$I_a = i_{l.a} - 0.62 \cdot n \cdot F \cdot A \cdot \nu^{1/6} \cdot \omega^{1/2} \cdot {D_R}^{2/3} \cdot c_{O,R} \tag{3.7}$$

where D_O and D_R are the diffusion coefficients of the reduced and the oxidized form, of the depolarizer, respectively, and $i_{l,c}$, $i_{l,a}$ are the cathodic and anodic limiting diffusion current, respecively.

Combining eqs. 3.5–3.7 yields the expression for the anodic–cathodic *I–E* curve[38]:

$$E = E_O + RT/nF \cdot \ln {D_R}^{2/3}/{D_O}^{2/3} + RT/nF \ \ln(i_{l,c} - i)/(i - i_{l,a}) \tag{3.8}$$

If the solution contains only one form of depolarizers, then either the anodic or the cathodic limiting current will be zero. Thus, following relation is derived from the cathodic *I–E*:

$$E = E_O + RT/nF \cdot \ln {D_{\text{red}}}^{2/3}/{D_{\text{ox}}}^{2/3} + RT/nF \ \ln(i_{l,k} - i)/\ i \tag{3.9}$$

And for the anodic *I–E* curve

$$E = E_O + RT/nF \cdot \ln {D_{\text{red}}}^{2/3}/{D_{\text{ox}}}^{2/3} - RT/nF \ \ln(i_{l,a} - i)/i \tag{3.10}$$

Since the diffusion coefficients of both forms of depolarizer are nearly identical, these equations can be simplified to:

$$E = E_O \pm RT/nF \ \ln(i_l - i)/i \tag{3.11}$$

If the radius of the electrode is small in comparison to that of the surrounding glass jacket, the access of solution to the electrode is uniform. The mass transport to the RDE can be expressed by the following equation:

$$\delta c_i/\delta t = D \ \delta^2 c_i/\delta x^2 + V_x \, \delta c_i/\delta x \tag{3.12}$$

The first term of this equation takes into consideration the diffusion of species *i* and the second term takes into account the convective transport of species *i* perpendicular to the disk. V_x is the velocity of solution perpendicular to the plane of the electrode, it is proportional to $\omega^{3/2}x^2$. ω is the rotation rate of the disk. Levich[39] has shown that convection is not much important when it is close to the electrode. Therefore in front of the electrode a boundary layer of thickness δ can be defined in which the mass transport is diffusion controlled. This layer is called Nernst diffusion layer (Fig. 3.8).

38 V. G. Levich"Physicochemical Hydrodynamics", Prentice-Hall Inc., Eaglewood Cliffs, N.J., 1962, p. 286ff.

39 V. G. Levich, "Physicochemical Hydrodynamics", Prentice-Hall Inc., Eaglewood Cliffs, N.J., 1962, p. 69.

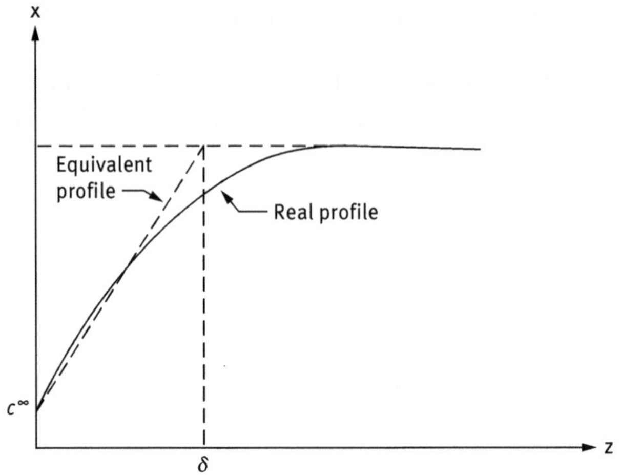

Fig. 3.8: Concentration profile of the electroactive species using the diffusion layer model of Nernst for the transport to the rotating disk electrode; δ is the thickness of the Nernst diffusion layer.

The thickness of the diffusion layer in front of a RDE is determined by the rotation rate of the disk, the extension of the layer shrinks with increasing rotation rate. The steady state current is given as follows:

$$I = -nFD\,(c_O(O) - c_{bulk}(O))/\delta = -\;nFD\,((dc(O)/dx)_{x=0} \qquad (3.13)$$

The current I depends on the square root of the rotational speed ω of the electrode. Thus the plot of the limiting current versus the square root of the rotation rate is linear as seen in Fig. 3.9.

From the relation between limiting current and $\omega^{1/2}$ one can obtain the diffusion coefficients D_O and D_R. The linearity is a proof of correct working of the electrode. Deviations from the linearity at high rotation rates point to the influence of turbulent mass transport when the Nernst layer becomes very thin. More information can be found in [40].

Outside this "diffusion layer" the concentration of all species is uniform and at their bulk values.

There are limits in using the RDE. The rotation rate ω should be in the range 100 rpm $< f <$ 10,000 rpm (10 s$^{-1} < \omega <$ 1,000 s^{-1}) provided the electrode surface is very finely polished and there are no eccentricities in the RDE shaft. At very high rotation rates there is no laminar flow but there is presence of turbulent flow from the bulk to the electrode. Then the relations given above are not valid.

40 A. J. Bard, L. R. Faulkner, "Electrochemical Methods", J. Wiley & Sons, New York 1980, p. 283ff.

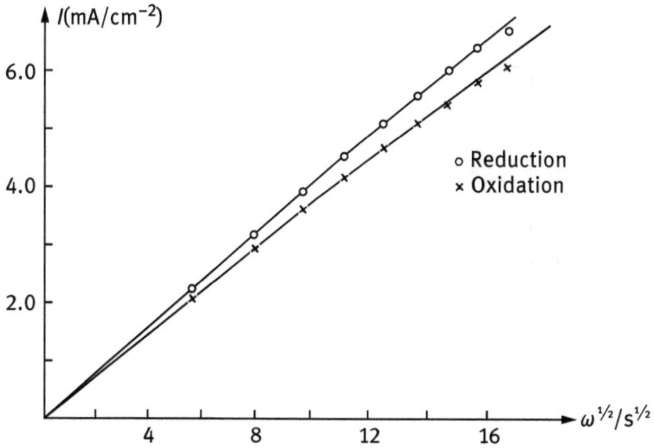

Fig. 3.9: Plot of the limiting current density i versus the square root $\omega^{1/2}$ of the rotation rate. The different slopes for the anodic and cathodic limiting current results from the different diffusion coefficients of the species O and R.

Additional problems may arise if the walls of the cell are too close to the electrode surface. The diffusion layer may be disturbed by the repulsion of the outstreaming flux.

The steady state profile of a solution containing the oxidized form O and the reduced form R of a depolarizer, for example, ferrocyanide/ferricyanide is illustrated in Fig. 3.10. where the complete I–E characteristic is depicted The solution contains $c_R{}^{\infty} = 3\, c_O{}^{\infty}$.

By slowly changing the potential of the working electrode, diffusion controls the electrode reaction and one obtains S-shaped *I–E* curves. Figure 3.10 shows three distinct zones labeled a–c.

Figure 3.10a shows the case of reversible electron transfer, for example, the charge transfer is very fast and the reaction is controlled by mass transfer. Starting from the left side of the diagram we are in the c section where the cathodic diffusion limited current $I_c = n\,F\,D_O\,c_O{}^{\infty}/\delta$ because of the reduction of O is observed. Note that this is a decreasing cathodic current that may be mixed with an increasing anodic current. At the intersection of the I–E curve with the potential axis at $I = 0$, we find the equilibrium potential E_e, where cathodic and anodic current compensate each other. In the following section, the current increases with increasing potential and finally reaches the plateau of the diffusion limited anodic current, $I_a = n\,F\,D_R\,c_R{}^{\infty}/\delta$.

Figure 3.10b shows the different shapes of an irreversible electrode process. Again starts the I–E curve in section c with the cathodic limiting current. However, the shape of the curve in the sections b and a is markedly different from that in Fig. 3.10a. The region around the equilibrium potential E_e corresponds to a current density less than a few percent of the limiting current density. Since the current

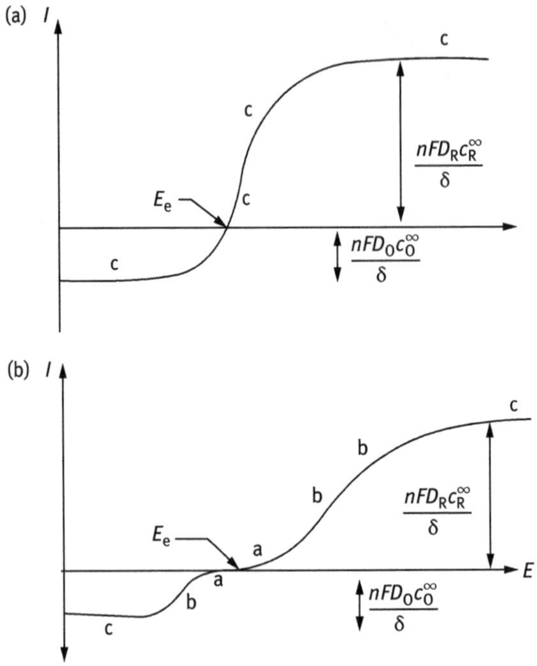

Fig. 3.10: Complete I–E curves over a wide range of overpotential for the reaction O + ne⁻ → R, where the solution contains $c_{bulk}(R) = 3\ c_{bulk}(O)$. (a) pure electron transfer control, (b) mixed control, (c) mass transfer control.

density is small, the range of the resulting chemical change at the electrode surface is small, and hence the surface concentrations of the species O and R are not significantly different from the bulk concentrations. Mass transport conditions will therefore have a negligible effect on the surface concentrations and therefore on the experimental current. Under these conditions the current is solely determined by the rate of the electron transfer and the equation as follows:

$$I = I_0\,\{\exp(\alpha_A nF/RT)\eta - \exp(\alpha_C nF/RT)\eta\} \tag{3.14}$$

This applies for very low values of η that may be expanded as series, and ignoring quadratic and higher order terms leads to the simple equation:

$$I = I_0(nF/RT)\eta \tag{3.15}$$

In region b, we find the intermediate situation of mixed control of the current by the mass transport and the electron transfer steps.

In region c the electrode process is again governed by mass transfer.

Whether an electrode reaction appears reversible or irreversible depends on both the kinetics of the electron transfer (standard rate constant k^0) and the mass transfer conditions. As a guide for the steady state experiment in unstirred solutions:

$$k^{\mathrm{o}} > 2 \cdot 10^{-2}\ \mathrm{cm\,s^{-1}}\text{leads to a reversible } I - E \text{ curve}$$

$$k^{\mathrm{o}} < 5 \cdot 10^{-3}\ \mathrm{cm\,s^{-1}}\text{ leads to an irreversible } I - E \text{ curve}$$

Figure 3.11 compiles the results given above and compares it with a Tafel plot.

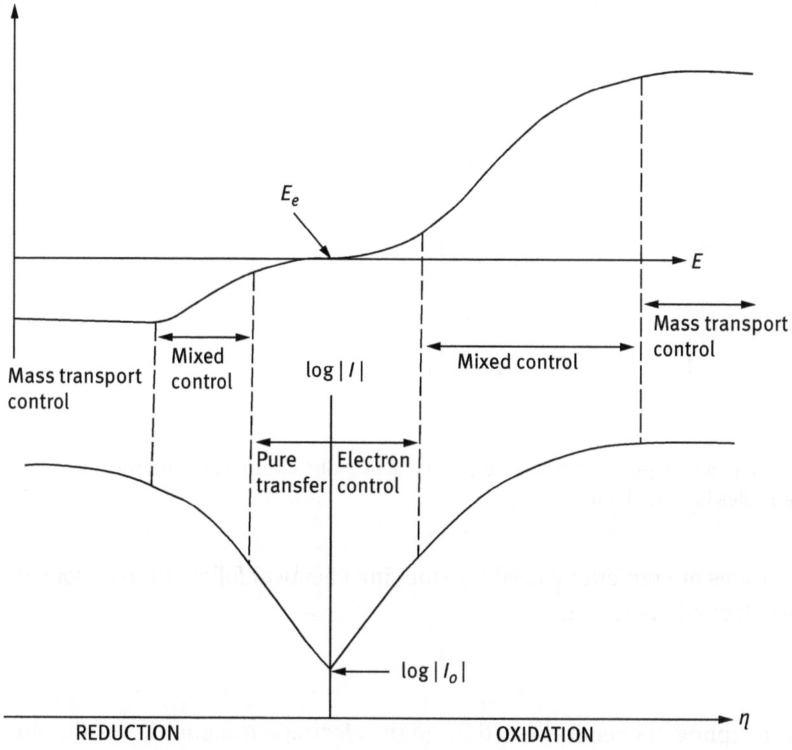

Fig. 3.11: I–E curve and corresponding log I–E curve for an irreversible electron transfer reaction. Solution as in Fig. 3.10.

In Figure 3.11, an extended range of the overpotential is shown. Only in the region a is the log I–E relationship linear, that is, only in this region Tafel plots can be measured. At the intersection of the linear Tafel plots the exchange current I_0 is obtained. For both the cathodic and the anodic processes the exchange currents are identical, they are compensated to zero at the equilibrium potential E_e.

It should be mentioned that the situation in practice may be more complex. The thickness of the diffusion layer of the RDE depends on the rotation of the electrode.

Increasing the rotation rates of the disk electrode lead to decreasing thickness of the Nernst diffusion layer and consequently to an increasing slope of the concentration profile of the electroactive species and because of this it leads to an increasing rate of the mass transport.

Finally, a short remark will be made on further developments of the rotating disk electrode. These are the rotation ring-disk electrode (RRDE) and the rotating double ring electrode (RDRE). A scheme of these electrodes is shown in Fig. 3.12. The electrodes are made of two concentrically electroactive areas that are separated by an isolating material. At the two electrodes, potential and current can be controlled, respectively and can be measured independently.

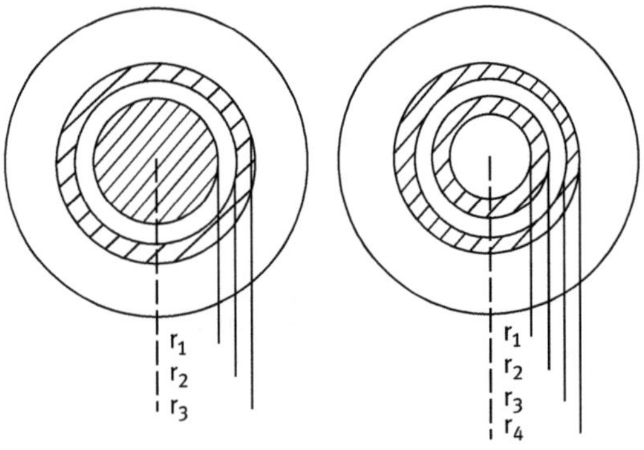

Fig. 3.12: Scheme of a ring-disk electrode (left) and of a ring-ring electrode (right); the thickness of the ring electrodes is about 1 mm.

These electrodes are preferably used for studying chemical follow-up reactions that follows the electrode potential:

$$R^- + A \rightarrow B^-$$

Generally, coupling of chemical reactions to the electrode reaction makes the situation more complex. It will be described in the following section on CV.

3.4.3 Cyclic voltammetry

Cyclic voltammetry (CV) is a potentiodynamic method, which means the working electrode potential is ramped linearly versus time The rate of the voltage change over time is known as the sweep rate or scan rate (V/s) of the experiment. The simplest potential-time waveform is sweeping the potential of the working electrode linearly between E_1 and E_2 at a known sweep rate v (V/s). This technique is called linear sweep voltammetry (LSV).

More used is the CV. In this technique the linear voltage sweep is not terminated at E_2 but reversed, usually at the same scan rate, to the initial potential E_1. Alternatively we can go to a value E_3. The cell current is recorded as a function of the applied potential. An important remark is at this point: the potential axis is also a time axis.

The cycles of ramps may be repeated as many times as needed. The solution is not stirred. The current at the working electrode is plotted versus the applied potential. The working electrode shows the typical I–E response of a cyclic voltammogram (Fig. 3.13).

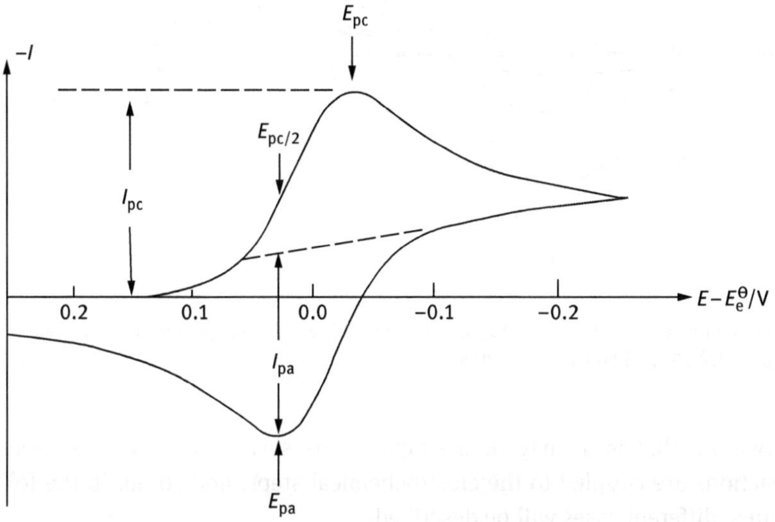

Fig. 3.13: A typical cyclic voltammogram of a reversible redox system O + e⁻ —> R.

The peak potentials E_{pc} and E_{pa} and the peak currents I_{pc} and I_{pa} characterize the voltammogram. The voltammogram depicted in Fig. 3.13 starts at a positive potential, the solution contains a redox couple (and excess supporting electrolyte). The observed current increases until the cathodic peak potential is reached. After this potential the current levels down until the potential E_{rev} is reached where the potential sweep is reversed. In the range $E < E_{pa}$ the electrode reaction is governed by the rate of electron transfer, whereas in the range $E > E_{pa}$ the electrode process is governed by the rate of mass transfer, that is, diffusion. The decay of the current images the growth of the diffusion layer in front of the electrode surface. The potential axis is simultaneously a time axis. The potential of reverse can be chosen arbitrarily. (Of course, the solvent and the supporting electrolyte must be stable at E_{rev} and the mass transport should not be influenced by natural convection.) In the reverse voltage sweep, we observe the reaction of the product generated in the upward sweep. We observe a cathodic current characterized by E_{pc} and i_{pc}. The shape of a cyclic voltammogram (CV) depends considerably on the nature of the system

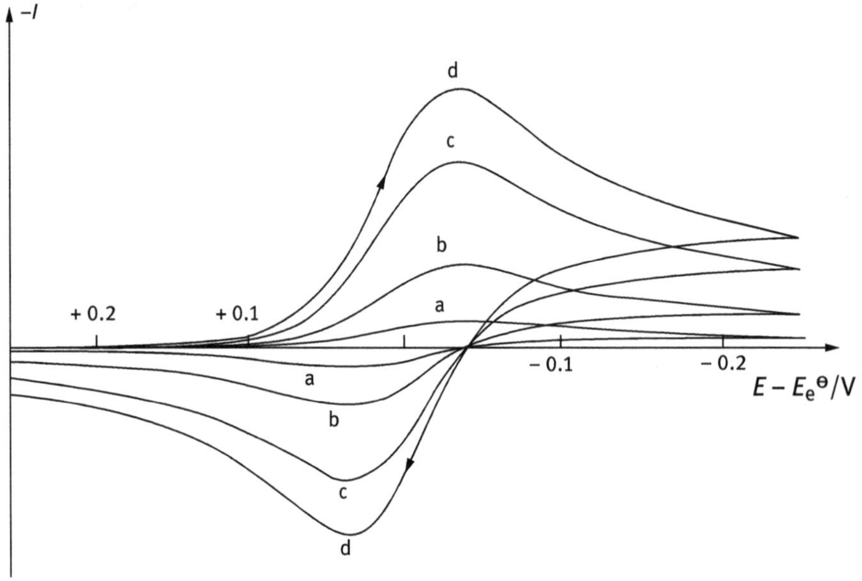

Fig. 3.14: The dependence of cyclic voltammogram of a reversible redox couple on the potential sweep rate *v*. (a) *v*, (b) 10 *v*, (c) 50 *v* and (d) 100 *v*.

under observation, that is, it may be a simple redox system or a system where chemical reactions are coupled to the electrochemical steps, and so on. In the following sections, different cases will be described.

3.4.3.1 Reversible couples

A CV of a reversible couple is shown in Fig. 3.13. It is observed when all the initial depolarizer can be recovered after a forward and reverse scan cycle. The wave form of the CV results from the combined effects of polarization and diffusion. The difference $\triangle E_p$ between the two peaks is of particular interest.

$$\Delta E_p = |E_{pc} - E_{pa}| \tag{3.16}$$

$\triangle E_p$ is $59/n$ mV (n = number of electrons involved in the redox couple). Another relation is the ratio I_{pa}/I_{pc}; for reversible redox systems it should be 1 (or very close to 1).

The exact form of the cyclic voltammogram can be determined by solving the Fick's second law:

$$\partial c_O/\partial t = D_O \partial^2 c_O/\partial x^2 \tag{3.17a}$$

$$\partial c_R/\partial t = D_R \partial^2 c_R/\partial x^2 \tag{3.17b}$$

For O and R with the appropriate initial and boundary conditions.

With O initially present in solution and assuming $D_O = D_R = D$, these initial and boundary conditions are as follows:

$$t = 0,\ x > 0,\ c_O = c_O{}^{\infty},\ c_R = 0$$

$$t > 0,\ x = \infty,\ c_O = c_O{}^{\infty},\ c_R = 0$$

$$t > 0,\ x = 0,\ D\ (\partial c_O/\partial x) + D(\partial c_R/\partial x) = 0$$

$$(c_O/c_R)_{x=0} = \exp[nF/RT(E - E_e{}^0)]$$

$$-I = n\ F\ D\ (\delta c_O/\delta x)_{x=0}$$

For a sweep rate v

$$0 < t < \lambda \qquad E = E_1 - vt$$

$$t > \lambda \quad E = E_1 - 2v\lambda + vt$$

E_1 is the initial potential and λ is the time at which the sweep is returned.

The solution of these equations[41] is not easy (see appendix). The peak current density I_p is given by the Randle–Sevcik eq. (3.16) for planar diffusion:

$$I_p = -0.4463\ n\ F\ (nF/RT)^{1/2} c_O{}^{\infty} D^{1/2} v^{1/2} \tag{3.18}$$

At 25 °C the equation reduces to the following form:

$$I_p = -(2.69 \times 10^5)\ n^{3/2} c_O{}^{\infty} D^{1/2} v^{1/2} \tag{3.19}$$

The meaning of the symbols and their dimensions is as follows:

Peak current density I_p (A cm^{-2}), diffusion coeficient D (cm^2 s^{-1}) and sweep rate v (V s^{-1}),$c_O{}^{\infty}$ (mol cm^{-3}). The peak current density is measured as shown in Fig. 3.13. The negative sign in eqs. (3.16) and (3.17) arises because of the condition that only O is present at $t = 0$, thus the peak current is a cathodic current.

The peak current density is proportional to the concentration of the electroactive species and to the square root of the sweep rate and diffusion coefficient.

Table 3.3 compiles the diagnostic criteria of the CV of a reversible system.
It has to be emphasized that a reversible system exists if all the criteria given in Table 3.3 are fulfilled. A failure to satisfy one or more of the conditions in Table 3.3 implies that the electron transfer is not reversible on the time scale of the experiment and that the process is more complicated than had been assumed.

The shapes of the cyclic voltammograms change when the rate of the electron transfer is insufficient to maintain the the Nernst equilibrium at the electrode surface.

41 J.E.B. Randles Trans. Faraday Soc., 44 (1948) 327; A. Sevcik, Czech. Chem. Commun. 13(1958); R.S. Nicholson and I. Shain, Anal. Chem., 36(1964) 706.

Table 3.3: Compilation of diagnostic tests for the voltammograms of reversible processes*.

1.	$\Delta E_p = E_{pa} - E_{pc} = 59/n \mathrm{mV}$
2.	$\lvert E_p - E_{p/2} \rvert = 59/n \mathrm{mV}$
3.	$\lvert I_{pa}/I_{pc} \rvert = 1$
4.	$I_p \propto v^{1/2}$
5.	E_p is independent of v
6.	At potentials beyond E_p $I^{-2} \propto t$

* This and the following tables are taken from "Instrumental Methodes in Electrochemistry" Ellis Horwood Limited 1985, p. 185ff.

At low scan rates of the potential the rate of electron transfer is still greater than that of mass transfer; a reversible CV is recorded. With increasing scan rate the rate of mass transport increases and becomes comparable to that of electron transfer. This leads to an increase in the peak separation. To analyze this effect it is recommended to plot $I/v^{1/2}$ versus the potential E. $I/v^{1/2}$ is the current normalized for the change in the rate of diffusion. For reversible systems the voltammograms superimpose at all sweep rates.

3.4.3.2 Irreversible couples

For irreversible systems the separation of peaks increases with increasing sweep rates and in addition the peak height is reduced in comparison to that of reversible systems.

The shape of the cyclic voltammogramm can be obtained by solving the differential eqs. (3.15a, b). The initial and boundary conditions remain the same as for the reversible case, except for $t > 0$, $x = 0$ that becomes

$$-I/n\,F = D_O\ (\partial c_O/\partial x)_{x=0} = \bar{k}\ (c_O)_{x=0} \tag{3.20}$$

Notice that at $t = 0$ only the species O is present and therefore the current density I is negative. The solution of this equation yields the following eq. (3.21) for the peak current density:

$$I_p = -\,(2.99 \times 10^5)\ n\ (\alpha_c n_\alpha)^{1/2}\ c_O{}^{\infty} D_O{}^{1/2}\ v^{1/2} \tag{3.21}$$

n_α is the number of transferred electrons including the rate determining step. The peak current density I_p (A/cm^2) of the irreversible system is proportional to the concentration and the square root of the sweep rate and in addition to the square root of the transfer coefficient α_c. The value of the transfer coefficient is often close to

0.5. With α_c equal to 0.5 the peak current for the irreversible process is only 78.5% in comparison to that of the reversible process.

For a deeper analysis of cyclic voltammogramms one can use working curves as shown in Fig. 3.16. Such working curves are available from the literature based on cyclic voltammogramms of other systems. In many cases, one obtains these by using kinetic parameters of coupled chemical reactions.

The most striking feature of a CV of a totally irreversible system is the complete absence of a reverse peak. However, this could also be because of a fast chemical reaction that consumes the product of the electron transfer. In contrast to the reversible system, where the peak potential is independent from the sweep rate the peak potential of the irreversible system shifts per decade change in the sweep rate v by $30/\alpha_c n_\alpha$ mV in a negative direction if v increases.

The shape factor $E_p - E_{p/2}$ is also different from the reversible case:

$$|E_p - E_{p/2}| = 48/\alpha_c n_\alpha \text{ mV at } 25°\text{C} \tag{3.22}$$

This equation permits the estimation of the factor $\alpha_c n_\alpha$ required for the estimation of D and k from eqs. (3.18) and (3.19).

Diagnostic tests for the CV of totally irreversible systems are compiled in Table 3.4.

Table 3.4: Diagnostic tests for a totally irreversible system at $T = 25$ °C.

1.	No reverse peak
2.	$I_p \alpha v^{1/2}$
3.	E_{pc} shifts −30 mV per decade increase in v
4.	$\|E_p - E_{p/2}\| = 48/\alpha_c n_\alpha$ mV

Reversible and totally irreversible systems are limiting cases; in between these systems quasi-reversible systems are also known. It is often observed that a system behaves reversibly at low sweep rates of the voltage and becomes irreversible at high sweep rates. This transition of a reversible system occurs when the relative rate of electron transfer with respect to that of mass transfer is insufficient to maintain Nernst equilibrium at the electrode surface. In the quasi-reversible region, both forward and backward reaction contribute to the observed current (Table 3.5). Plotting I_p versus $v^{1/2}$ as depicted in Fig. 3.15 readily shows the quasi-reversible region. This region is generally recognized to have the following boundaries[42]:

42 H. Matsuda, Y. Ayaabe, Z. Elekrochemie, 59(1955) 494.

Table 3.5: Diagnostic tests for quasi-reversible systems.

1.	$\lvert I_p \rvert$ increases with $v^{1/2}$ but is not proportional to it
2.	$\lvert I_p^A / I_p^C \rvert = 1$ provided $\alpha_C = \alpha_A = 0.5$
3.	ΔE_p is greater than 59/n mV and increases with increasing v
4.	E_p^C shifts negatively with increasing sweep rate v

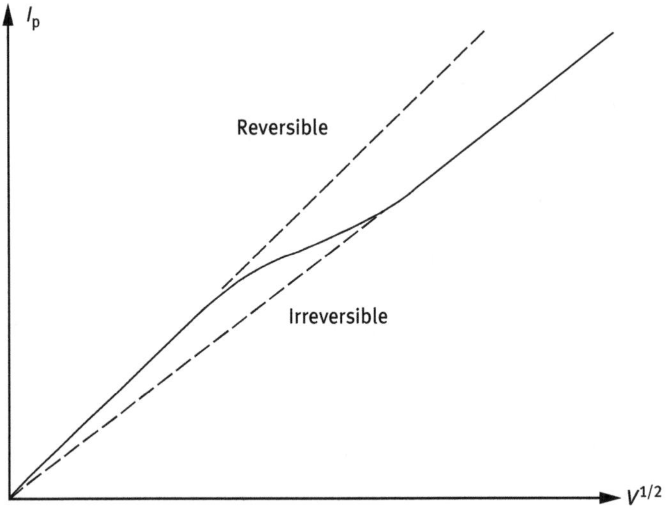

Fig. 3.15: The peak current plotted versus the square root of the potential sweep rate, showing the transition from reversible to irreversible behavior with increasing sweep rates.

$$0.3\, v^{1/2} > k^0 > 2 \times 10^{-5} v^{1/2} \mathrm{cm\, s^{-1}}$$

Finally, it should be mentioned that kinetic data, such as k^0, are often obtained by using working curves,[43] which have been constructed for $n \Delta E_p$ as a function of the variable ψ defined as follows:

$$\psi = (RT)^{1/2} k^0 / (n\, F\, D \pi\, v)^{1/2} \tag{3.23}$$

As an example, Fig. 3.16 shows a working curve where $\triangle E_p$ is plotted versus ψ. Comparing the experimental E_p with the working curve for several v values k^0 is obtained.

43 R.S. Nicholson, Anal. Chem., 37 (1965) 1351.

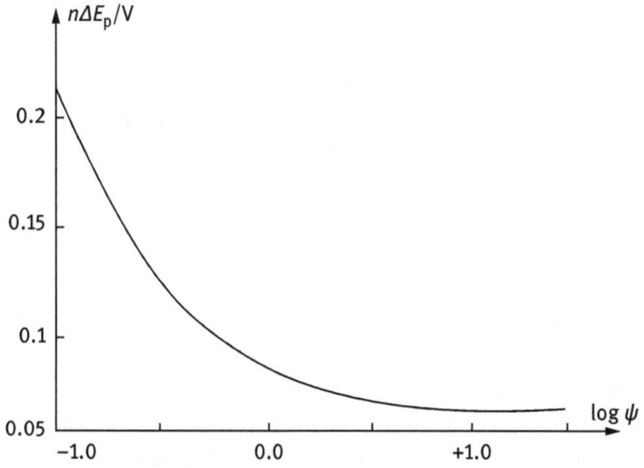

Fig. 3.16: The working curve of $n\Delta E_p$ versus log ψ ($\psi = (RT/nFD\pi v)^{1/2}k^0$) for the determination of the standard rate constant for electron transfer from the separation of anodic and cathodic peaks in a cyclic voltammogram.

3.4.4 Studying chemical reactions

CV is probably the most powerful electrochemical technique for studying coupled chemical reactions and in this way detecting and identifying reaction intermediates and evaluating the complete mechanisms occurring in an electrochemical system. In the classification of mechanisms stands E for electron transfer and C for a chemical reaction. Thus in CE, a chemical reaction is preceding the electron transfer, ECE symbolizes a chemical reaction between two electron transfer processes. This type of reaction allows the characterization of the product of the first electron transfer. Another often observed type is the EC mechanism.

3.4.4.1 The CE system

The CE reaction can be symbolized as follows:

$$Y \underset{k_{-1}}{\overset{k_1}{\rightleftharpoons}} O$$

$$O + ne^- \rightleftharpoons R$$

Typical examples are weak acids and the reduction of formaldehyde where an electroactive form and an electroinactive hydrated form exist.

$$HCH = O + H_2O \rightleftharpoons H_2C{=}C(OH)_2$$

$$H_2C = O + e^- \rightleftharpoons H_2C{=}O^-$$

The appearance of the CV depends on the nature of the preceding chemical reaction, whether or not it is unimolecular or bimolecular and in addition it depends on the rate of reestablishing a disturbed equilibrium. Also the nature of the electron transfer strongly influences the form of the CV.

We first describe a system with very slow preceding chemical reaction and reversible electron transfer. The current will be completely controlled by the kinetics of the chemical reaction. No diffusion control is involved in the process. Therefore no peak currents are observed in cyclic voltammogramm. A simple steady state is observed. The rate constants of the chemical reaction can be obtained directly from the limiting current[44] using the following equation:

$$I_L = -n\ F\ c_y^{\infty}\ D\ K\ (k_1 + k_{-1})^{1/2} \tag{3.24}$$

K is the equilibrium constant, which has to be determined.

On the other side, if the chemical reaction is extremely fast, the cyclic voltammogramm images a simple diffusion controlled electron transfer. Under intermediate conditions the surface concentration of the reducible species O and hence the current will be partially controlled by the kinetics of the chemical reaction and the shape of the cyclic voltammogramm will be slightly different from that for reversible behavior, particularly that of the forward peak. The transition from purely kinetic control to purely mass transport control is a complex function of the relative values of K, k_1, k_{-1}, and the voltage sweep rate v. The reader can refer to the publications of Saveant and Vianello[45].

Diagnostic tests for the CE mechanism:
$I_{pc}/v^{1/2}$ decreases as v increases
$|I_{pa}/I_{pc}$ increases with v and is always greater or equal tounity

---

3.4.4.2 The EC system

The reacting system is given as follows:

$$O + ne^- \rightarrow R$$

$$R \underset{k_{-1}}{\overset{k_1}{\rightleftharpoons}} Y$$

The EC scheme is a very common mechanism in organic electrochemistry. The cyclic voltammogramm observed depends on the relative rates of the two steps. A simple situation is where electron transfer is totally irreversible; the presence of the chemical reaction has no effect on the voltammogramm obtained and no kinetic data related to the chemical reaction can be derived (Table 3.6).

44 J.M. Saveant and E. Vianello, Electrochim. Acta, 8 (1963)905, 12 (1967) 629, 1545.
45 L. Nadjo and J.M Saveant, J. Electroanalyt. Chem., 48 (1973) 113.

Table 3.6: Diagnostic criteria for the EC reaction.

1.	I_{pa}/I_{pc} I is less than 1 but tends to unite as v is increased
2.	$I_{pc}/v^{1/2}$ decreases slightly with increasing v
3.	E_{pc} is positive from the value for the reversible case
4.	E_{pc} shifts negatively with increasing v, and i is the pure kinetic region that shifts by 30/n mV per tenfold increase in v for first-order reaction and 19/n mV for a second-order reaction.

The full range of possibilities, where the chemical reaction can be reversible or irreversible and the electron transfer that is either reversible or quasi-reversible has been published in detail by Nadjo and Saveant and Nicholson and Shain.

In the following sections, only the case is described where the electron transfer is reversible and the chemical reaction is irreversible.

The effect of a chemical reaction following the electron transfer is of course greatest on the reverse sweep. Using a low sweep rate and the fast chemical reaction, it leads to a fast depletion of R near the electrode. No reverse peak will be observed. If the time scale of the experiment is lowered by increasing the sweep rate, the reverse peak will increasingly be observable. As an example, normalized cyclic voltammogramms are shown (fig. 3.17) in the case of reversible electron transfer and irreversible chemical reaction for a range of values of the variable λ given as follows:

$$\lambda = k\ (RT/nF)/v \tag{3.25}$$

In the range where no reverse peak is observed ("kinetic zone") the cathodic peak potential E_{pc} is shifted positive of the E_{pc} value for reversible electron transfer, because the coupled chemical reaction reduces the concentration of R at the electrode surface; from the values, it would have had a simple electron transfer reaction. However, the peak is shifted back to 30/n mV (at 25 °C) in a negative direction for each tenfold increase in the sweep rate v, provided the chemical reaction is of first order. If the chemical reaction is of second order in R, then $\partial E_{pc}/\partial \log v = -19.6$ mV. The ratio of the anodic to the cathodic peak current increases with increasing sweep rate until eventually reversible behavior is observed. In the relatively small range $5 > \lambda > 0.1$, where a reverse peak is observed but the ratio I_{pa}/I_{pc} is less and kinetic data can be obtained by comparing experimental I_{pa}/I_{pc} ratios with the working curve published by Nicholson and Shain.[46] In this chapter, this ratio is plotted as a function of log $k\tau$, where τ is the time required to traverse the potential range from the polarographic half wave potential $E_{1/2}$ to the switching potential E_2 at which the sweep is reversed. The value of the half wave potential $E_{1/2}$ can be obtained under reversible conditions from the difference $E_{pc} - E_{1/2}$:

46 R.S. Nicholson and I. Shain, Anal. Chem., 37 (1965), 178 and 190.

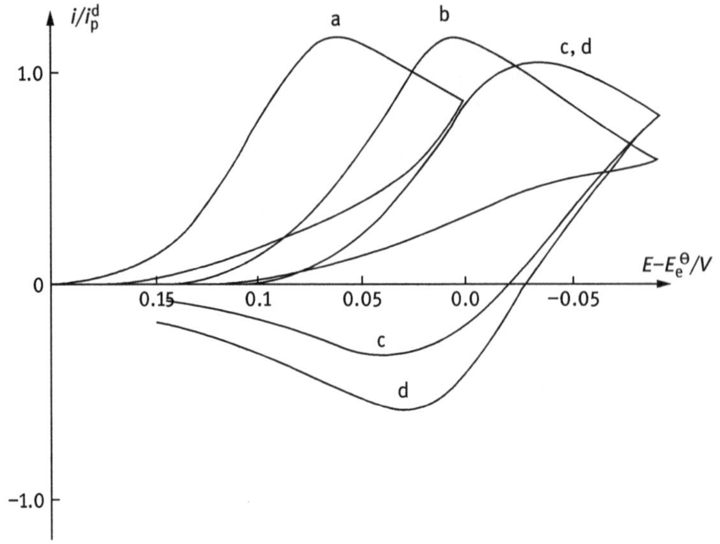

Fig. 3.17: Theoretical cyclic voltammogramms for the EC mechanism for the following values of λ: (a) 500, (b) 10, (c) 0.1, (d) 0.01.The current scale is normalized to the peak current in the absence of a chemical reaction.

$$E_{pc} - E_{1/2} = -1.109\ RT/nF = -28,5/n\ mV \text{ at } 25°\text{C} \tag{3.26}$$

To get reliable values of rate constant, a range of switching potentials and sweep rates should be used. An alternative approach based on peak potential measurements arises from the kinetic zone analysis by Nadjo and Saveant.[47]

The ECE mechanism

$$\mathrm{O} + \mathrm{n_1 e} \rightarrow \mathrm{R}$$

$$\mathrm{R} \xrightarrow{k} \mathrm{O'}$$

$$\mathrm{O'} + \mathrm{n_2 e} \rightarrow \mathrm{R'}$$

The ECE mechanism is frequently observed in the electrochemistry of organic compounds. It is a specific type of reaction where the reaction product R of the first electron transfer reacts chemically to yield a species O′ which itself is reduced to the species R′. This second electron transfer may occur at the same or in fact at a more positive potential. The appearance of the cyclic voltammogram depends on the nature and rate of the chemical reaction.

In the following sections, we restrict ourselves to the case where the electron transfer processes are reversible and the chemical reaction is an irreversible reaction.

47 L. Nadjo and J.M. Saveant, J. Electroanal. Chem., 48 (1973), 113.

First we consider the forward sweep. When the rate k of the chemical reaction is large compared to the rate of mass transport the electrode reaction will appear to be an (n_1+n_2) electron process, whereas if k is low it will behave like n_1 electron. Thus for any given system the apparent number n_{app} of electrons involved will decrease from (n_1+n_2) toward n_1 if the sweep rate is increased. Changes are also observed in the reverse and subsequent sweeps. Figure 3.18 shows a cyclo-voltammogram with the rate of mass transport comparable to k^0.

Peaks I and II correspond to the couple O/R whereas III and IV correspond to the couple O′/R′, thus for increasing values of k or decreasing sweep rate, peaks III and IV become increasingly apparent.

The behavior of an ECE system is most clearly seen by plotting $/I_{pc}\, v^{1/2}$ for peak I on the first sweep as a function of v or log v (analogous to the $It^{1/2}$ vs. t plots in chronoamperometry). At low sweep rates, mass transport is slow and therefore a peak corresponding to an irreversible two-electron process is seen, whereas at higher sweep rates only a reversible one-electron step is found. By analyzing such a $I_{pc}/v^{1/2}$ versus log y plots values of k can be found.
Nicholson and Shain have shown that

$$I_{pc}/v^{1/2} = -\pi^{1/2} F\, c_{=}^{\infty} (D\, nF/RT)^{1/2} (n_1 \chi c_O^{\infty} + n_2 \Phi c_O^{\infty}) \qquad (3.27)$$

where χ and Φ values are available in tabulated form in the original literature as functions of (kRT/nFv). In principle by a curve fitting of experimental values $I_{pc}v^{1/2}$ data to eq. (3.27) values of k can be determined.

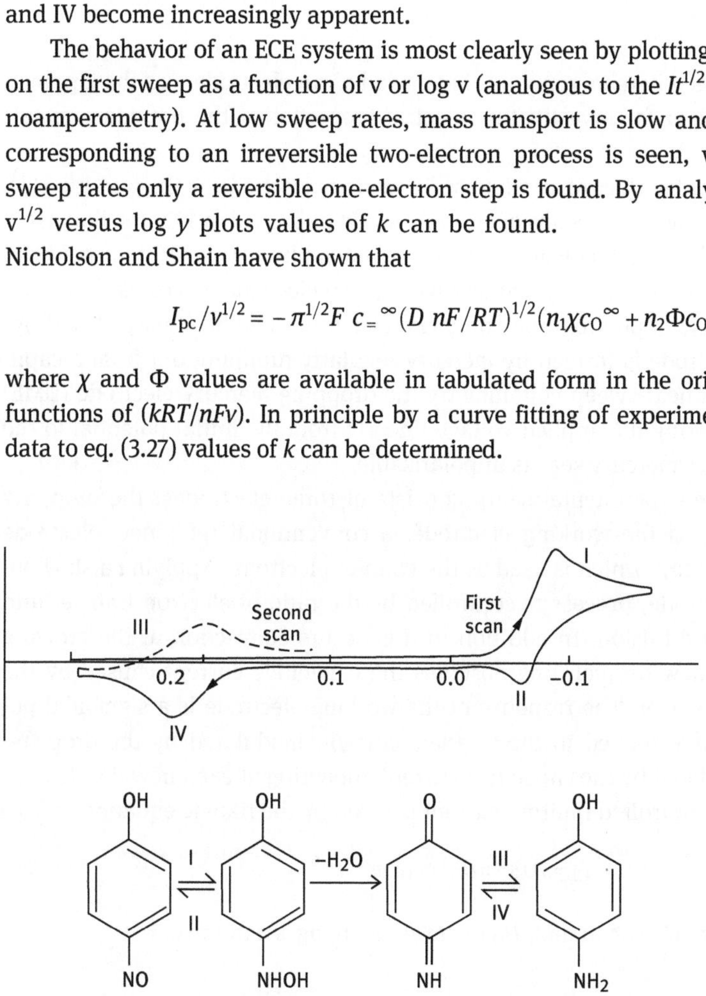

Fig. 3.18: Cyclic voltammogram for the reduction of p-nitrosophenol in acetate buffer at a mercury drop electrode. (Structures provided by https://pubchem.ncbi.nlm.nih.gov).

Table 3.7: Diagnostic criteria for the ECE reaction.

1.	$I_{pc}/v^{1/2}$ varies with the sweep rate but may reach limiting values but at high and low sweep rates; $I_{pc}/v^{1/2}$ (low v). > $I_{pc}v^{1/2}$ (high v)I
2.	I_{pa}/I_{pc} increases with sweep rate and tends to be 1 at high sweep rates

Information on more complex systems can be found in the literature.[48]

3.4.5 Polarography

The first instrument where mercury has been used as an electrode was the Lippman capillary electrometer for investigating electric properties of the boundary metal/electrolyte.

It was then Jaroslav Heyrovsky who used a mercury electrode as the working electrode to analyze solutions by means of electrolysis. He published in 1922 the first paper on polarography.[49] In the following decades, polarography developed to a powerful tool in electro-analytical chemistry and for investigating electrode processes.

In the early days of polarography only two electrodes had been used. The dropping mercury electrode is formed by mercury regularly dropping out from a capillary tube. Like in linear-sweep voltammetry the dropping mercury electrode (DME) is polarized by altering the applied voltage linearly from the initial potential to the final potential. The "mercury sea" is unpolarizable.

Nowadays, the experimental setup consists of three electrodes: the dropping mercury electrode as the working electrode, a conventional reference, electrode and the "mercury sea," which is used as the counter electrode. Applying a slow potential sweep electrode, process is controlled by the ratio of electron transfer and mass transport by diffusion. In addition to this occurs convection at the growing droplet. At every new droplet, one observes the capacitive current caused by the developing double layer. The response of the working electrode is a sigmoidal potential–current curve related to the faradaic current, modulated by the drop frequency and modulated by the capacitive current appearing at each new droplet.

The diffusion controlled limiting current is given by the Ilkovic equation[50]:

$$I_d = 708 \cdot nD_O{}^{1/2} c_O{}^{*} m^{2/3} t^{1/6} \tag{3.28}$$

where i_d is taken in μA, $c_O{}^{\star}$ in mM, D_O in cm^2/s, m in mg/s and t in s.

48 "Instrumental Methods in Electrochemistry", Ellis Horwood Ltd., Chichester 1985, p. 214f.
49 J. Heyrovsky, Chem. Listy, 16, 256 (1922).
50 D. Ilkovic, Coll. Czech. Chem. Commun., 6, 498 (1934).

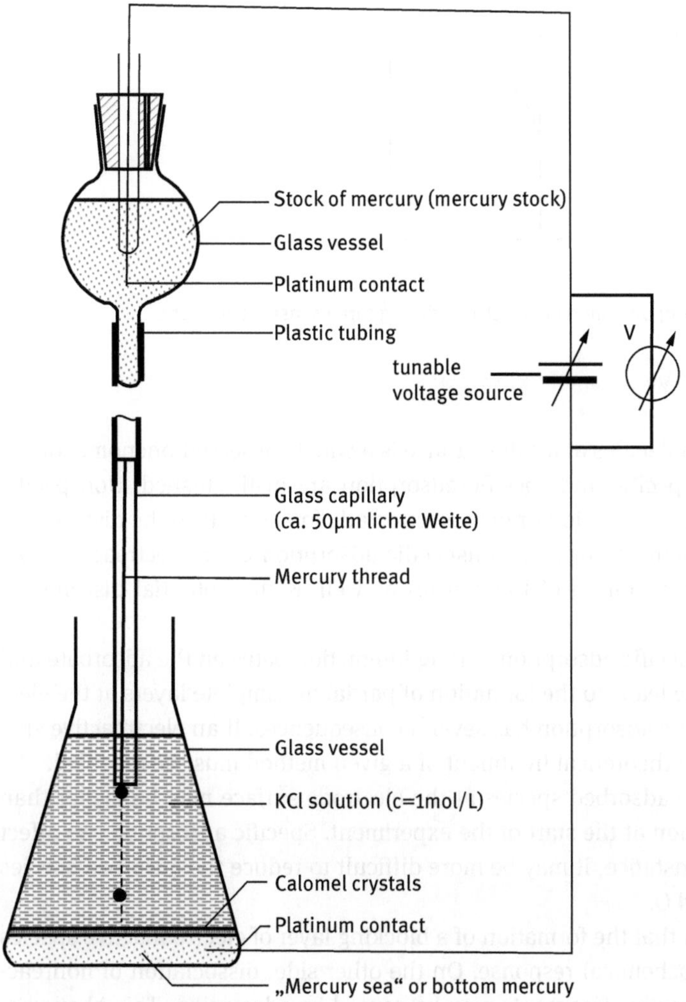

Fig. 3.19: Dropping mercury electrode. The simplest experimental setup is shown. It consists of the polarizable drop electrode, the unpolarizable "mercury sea" and variable voltage source. In a three electrode setup the "mercury sea" represents the counter electrode and conventional reference electrode is added.

Figure 3.20 shows the current growth during the successive single drops of a dropping mercury electrode.

A detailed description of polarography and its application can be found in the literature[51].

51 A.J. Bard, L.R. Faulkner, "Electrochemical Methodes" J. Wiley &Sons, New York 1980, p. 145ff.

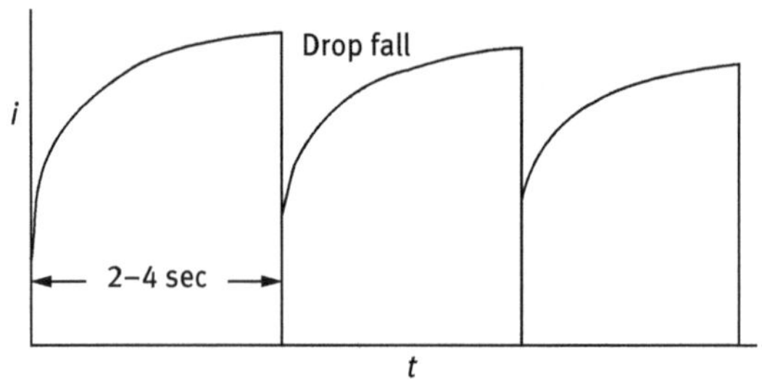

Fig. 3.20: Current at a dropping mercury electrode during three consecutive drops.

3.4.6 Cyclic voltammograms and adsorption

The adsorption of molecules at a solid surface is a widely observed phenomenon. In the literature, nonspecific and specific adsorption are distinguished. Nonspecific adsorption is the case where long-range electrostatic forces perturb the distribution of ions near the electrode surface. Nonspecific adsorption of an electroactive species affects the concentration of the species as well as the potential distribution near the electrode.

In the case of specific adsorption, strong interaction between the adsorbate and the electrode surface leads to the formation of partial or complete layers at the electrode surface. Specific adsorption has several consequences. If an electroactive species is adsorbed, the theoretical treatment of a given method must be modified. The concentration of the adsorbed species at the electrode surface may be higher than the bulk concentration at the start of the experiment. Specific adsorption can affect the energetics, for instance, it may be more difficult to reduce the adsorbed species O than the dissolved O.

It is well known that the formation of a blocking layer of electro inactive species can alter the electrochemical response. On the other side, dissociation of nonreactive material into reactive fragments may be caused by adsorption. This phenomenon is called electrocatalysis, because the electrode behaves as a catalyst.

In the following sections, we describe qualitatively the influence of adsorption on cyclic voltammograms. The quantitative treatment of adsorption can be found in literature.

We begin with a reversible system where O and R are adsorbed and where only the adsorbed species O is electroactive (fig. 3.21). This could be the case, where the sweep rate is so fast that no appreciable diffusion of O at the electrode surface occurs. There are also cases where the adsorption is so strong that the adsorbed layer of O can be formed even when the concentration is so small that the contribution to the current from dissolved O is negligible. In addition, we assume that within the range of potentials of the wave the excess concentrations Γ are independent from the potential.

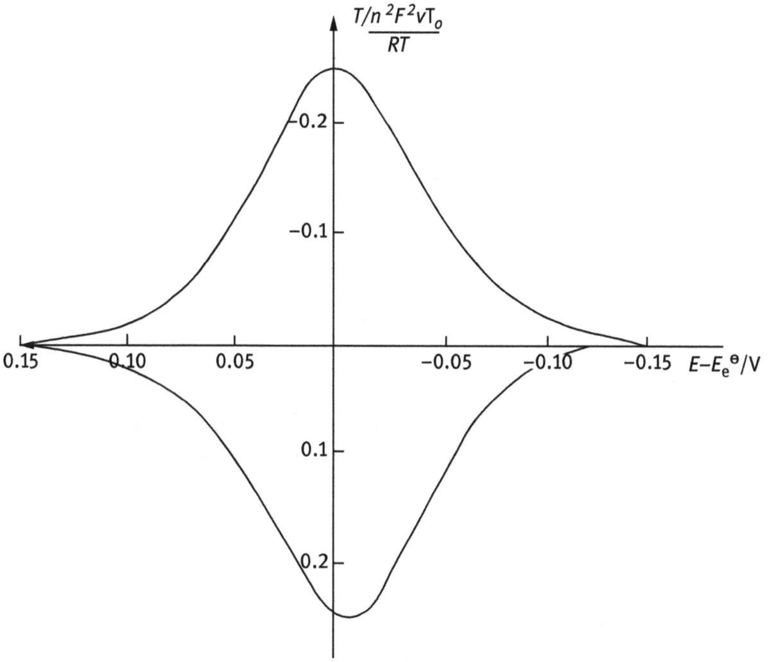

Fig. 3.21: Theoretical cyclic voltammogram for the reduction of adsorbed O and reoxidation of the product where the free energies of adsorption of O and R are equal.

The peak heights are given as follows:

$$I_p = (n^2F^2/4RT)\ v\ A\ \Gamma_O \tag{3.29}$$

with $\Gamma_{O*} = \Gamma_O(t) + \Gamma_R(t)$

and the peak potential is given as follows:

$$E_p = E^0 - (RT/nF) \cdot \ln(b_O/b_R) \tag{3.30}$$

with $b_O = \beta_O \Gamma_{O,s}$ and $b_R = \beta_R \Gamma_{R,s}$

The location of E_p with respect to $E^=$ depends on the relative strength of adsorption of O and R; if $b_= = b_R$ then $E_p = E^0$. If O is adsorbed more strongly ($b > b_R$) the wave is displaced toward more negative potentials beyond the position where the reversible wave of a diffusing species would occur. If on the other side $b < b_R$ the wave is shifted to more positive potentials than E^0. When both the dissolved and the adsorbed species are electroactive the theoretical treatment becomes rather complicated.

In the following sections, we describe qualitatively the case for a reversible electron transfer where either the reactant O (fig. 3.22) or the product R (fig. 3.23) is adsorbed, but not both.

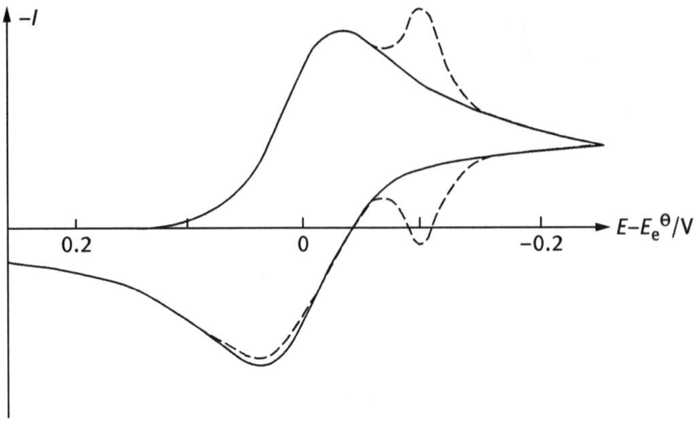

Fig. 3.22: Cyclic voltammogram when O is strongly adsorbed. The dashed line indicates the response with adsorption of O, the solid line that of a simple reversible system without adsorption.

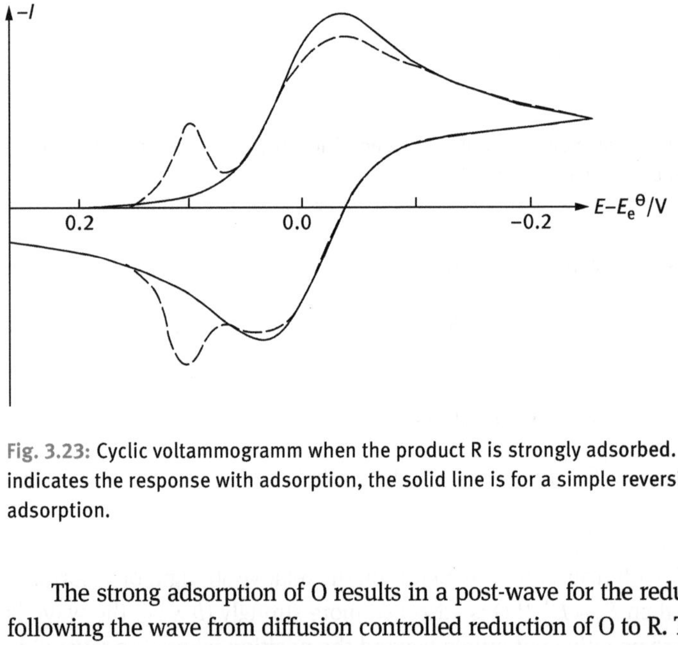

Fig. 3.23: Cyclic voltammogramm when the product R is strongly adsorbed. The dashed line indicates the response with adsorption, the solid line is for a simple reversible process without adsorption.

The strong adsorption of O results in a post-wave for the reduction of adsorbed O, following the wave from diffusion controlled reduction of O to R. The post-wave results from the greater stability for reduction of adsorbed O in comparison to dissolved O.

In a system where the product R is strongly adsorbed one observes a prewave, representing the reduction of dissolved O to form a layer of adsorbed R. The wave is observed at potentials that are more positive than the diffusion controlled wave, because the free energy of adsorption of R allows an easier reduction of O to adsorbed R than to dissolved R. The peak current of the prewave increases with the sweep rate v, whereas that of the diffusion peak varies with $v^{1/2}$.

3.4.7 Some final remarks

At this point, it is worth mentioning how a double layer charging and iR drop manifests in cyclic voltammograms.

As with all transient techniques, there is a contribution of double layer charging to the total current

$$I_{\text{total}} = I_{\text{Faraday}} + I_{\text{dl}} \tag{3.31}$$

The charging current i_{dl} of the double layer is as follows:

$$I_{\text{dl}} = C_{\text{dl}} \cdot \text{v} \tag{3.32}$$

The charging current I_{dl} is proportional to the sweep rate v, while the Faraday current is proportional to $\text{v}^{1/2}$. After a potential jump the decay of the capacitive current is much faster than that of the Faraday current.

The double layer capacity typically is in the range 20 to 40 μF/cm², at a sweep rate of 100 mV/s, the current density of double layer charging will be between 2 and 4 μA/cm²; in comparison to the Faraday current this is almost negligibly. However, at a sweep rate of 100 V/s the current density of double layer charging is between 2 and 4 mA/cm²; these values can no longer be neglected; in fact, they may be a limitation for using large scan rates in CV measurements. To overcome this barrier the use of microelectrodes is recommended.

Another problem is the *iR* drop, which changes the experimentally applied voltage E_{ap} from the desired voltage E.

$$E_{\text{ap}} = E - iR$$

The effects on the CV are as follows: decreasing peak heights and decreasing peak separation. These effects affect the analysis of the cyclo-voltammograms.

The best way to check the influence of iR drop is by measuring a CV of a reversible couple such as cobaltocene/cobalticinium or ferrocene/ferrocenium under identical conditions, that is, cell, electrodes, electrolyte, solvent, potential scan rate and concentration of the electroactive substance. From the deviation of the observed CV from the expected CV the influence of the *iR* drop can be evaluated.

4 Applications of electrochemistry

Electrochemistry is a branch of physical chemistry and is connected well with chemistry as well as with physics and there are many relations with other disciplines such as biology and medicine, geology and metallurgy and last but not the least engineering.

Analytical chemistry is essential for applying electrochemical methods. However, there is also a considerable number of processes where electrochemistry is used in industrial dimensions. The electrochemical production of chlorine, which is one of the major chemical products goes back to the late nineteenth century. Aluminum, which is one of the most utilized metals, is produced by electrolysis of a melt. The electrochemical synthesis of adiponitrile, a preproduct of nylon, can be performed by a large-scale electrochemical process.

The storage of electrical energy in rechargeable batteries and the use of fuel cells as primary sources of electrical energy is a key for electromobilization, and thus solves problems of air pollution. Electrochemical reactions are the basic processes of corrosion, the main cause for degradation of metallic constructions. On the other side, many ways of obtaining corrosion-resistant surfaces of metals are electrochemical in nature. Etching of silicon for the fabrication of chips and micromechanical devices is based on electrochemical processes. Finally, it can be mentioned that the transfer of signals through nerves is based on electrochemical mechanisms.

In the preceding chapters the fundamentals of all these applications have been presented. However, there are still more aspects of applications of electrochemistry that are worth describing and explaining. These are the objectives to which the following chapters are devoted to.

4.1 Electroanalytical techniques

There are many different possibilities to solve analytical problems by using electrochemical methods. The principal point is the evaluation of the concentration of specific species in solution. Often additional species are present demanding high specificity of the applied method.

In titrations, electrochemistry can be used to find the final point. In many methods of determination, the current in an electrode process contains the information of the concentration of a species. A specific case of analytical tools is electrochemical sensors, which indicate the concentrations of specific species.

In various fields, analytical applications of electrochemistry has become routine. Environmental research (water, atmosphere and soil), metallurgy, geology, pharmacy, clinical chemistry, food chemistry are typical examples. Often electrochemical methods are advantageous because measurements in colored or cloudy liquids are possible.

https://doi.org/10.1515/9783110437393-005

Another advantage is the possibility to determine simultaneously different components of a solution. The use of ion-selective electrodes simplifies measurements, so that field measurements are possible. Generally, electrochemical analytical instruments require less money in comparison to spectroscopic setups.

The great variety of methods and analytical problems make it difficult to give a complete overview of the field; therefore, only a selection of frequently used methods will be given in the following sections.

4.1.1 Applications of potential measurements

4.1.1.1 Standard electrode potentials

For the determination of standard electrode potential as per definition, an electrochemical cell is used consisting of a standard hydrogen electrode and the electrode of interest. For exact measurement, this electrode is frequently used as a metal M in its aqueous solution M^+X^-. For the determination of the standard potential of the cation and the anion ion activities have to be taken into account.

4.1.1.2 Determination of solution products

In the following sections, we take the silver chloride electrode as an example. We assume that the standard potential of the silver electrode is known; this simplifies the problem. The half cell is a silver wire coated with silver chloride and immersed in a solution saturated with silver chloride.

$$(\mathrm{AgCl})\mathrm{Cl}^-(\mathrm{c})/\mathrm{Ag}$$

The current-free equilibrium potential of this cell is given as follows:

$$E_{\mathrm{Ag}} = E^0{}_{\mathrm{Ag}} + RT/F \ \ln a(\mathrm{Ag}^+) \tag{4.1}$$

Since the solution is saturated with AgCl, the activity of the silver ion in the solution is given by the solubility product K_s of silver chloride:

$$K_s = \mathrm{a}(\mathrm{Ag}^+) \cdot \mathrm{a}(\mathrm{Cl}^-) \tag{4.2}$$

The equilibrium potential of the silver chloride electrode comes to the following:

$$E_{\mathrm{Ag}} = E^0{}_{\mathrm{Ag}} + RT/F \ \ln K_s - RT/F \ \ln a(\mathrm{Cl}^-) \tag{4.3}$$

From this equation follows:

$$\ln K_s = F/RT(E^0(\mathrm{Cl}^-)\,|\mathrm{AgCl}|\,\mathrm{Ag}) - E^0(\mathrm{Ag}^+\,|\mathrm{Ag}) \tag{4.4}$$

Using the standard potential of the AgCl electrode and that of the metal–ion electrode the solubility product K_s can be calculated.

4.1.2 Determination of the pH value

The pH_a value in aqueous solutions is defined by the negative decimal logarithm of the proton activity:

$$pH_a = -\log a(H^+{}_{aq}) \tag{4.5}$$

For the determination of the pH_a value one needs a galvanic cell, which contains at least one half cell that is sensitive for the activity of protons (fig. 4.1). Both the half cells are separated by a salt bridge or a membrane. The reference electrode may be one of the standard reference electrodes.

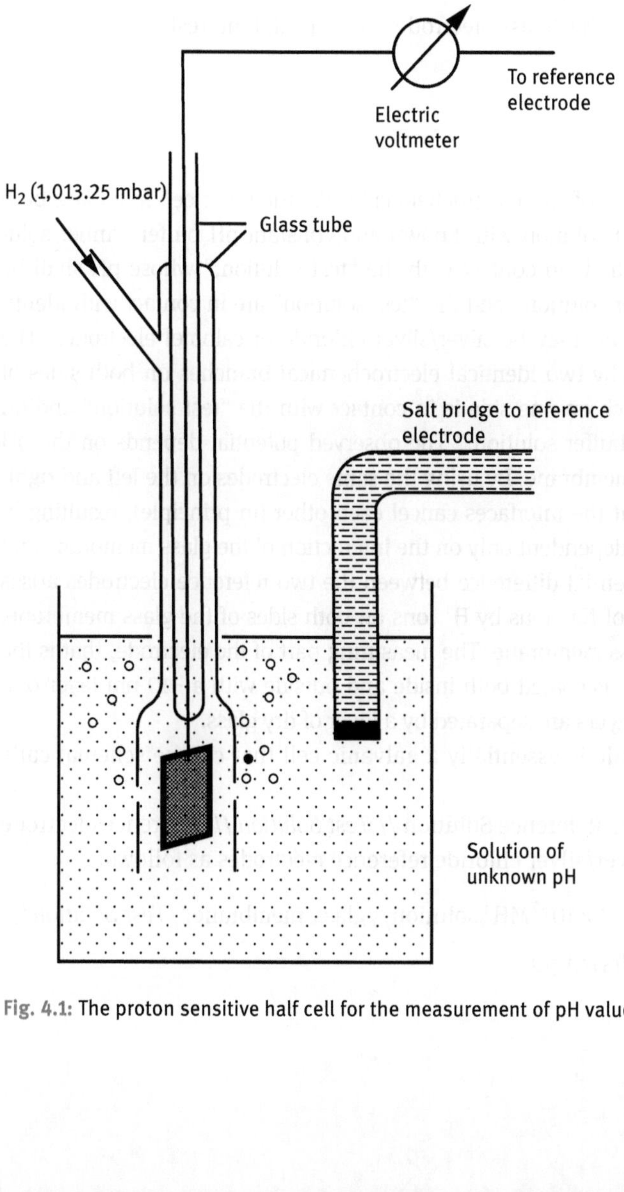

Fig. 4.1: The proton sensitive half cell for the measurement of pH value.

It is possible to use a hydrogen-concentration chain, where one hydrogen electrode is immersed in the solution with unknown proton concentration and the standard hydrogen electrode is used as the reference electrode. The potential E is given as follows:

$$E = RT/F\ (\ln\ aH^{+}{}_{aq}) + \Delta\varphi_{diff} \tag{4.6}$$

and the pH value by

$$pH_a = -(E - \Delta\varphi_{diff})\ F/RT\ \ln 10 \tag{4.7}$$

There are sometimes substances in the solution that contaminates the electrode, so that the hydrogen potential cannot be adjusted properly; if this is the case the quinhydrone electrode can be used.

For pH measurements the "glass electrode" is of special interest.

4.1.3 Glass electrode

The glass electrode consists of two electrochemical cells: the inner cell in a glass bulb with thin walls contains a solution with known and constant pH buffer ("inner solution"). The "inner solution" is in contact with the "test solution," whose pH shall be measured. Both the "inner solution" and the "test solution" are in contact with identical reference electrodes; this may be silver/silver chloride or calomel electrodes. The glass electrode is formed by two identical electrochemical branches on both sides of the glass membrane, which on one side is in contact with the "test solution" and on the other side with the "buffer solution." The observed potential depends on the pH gradient along the glass membrane. By using the same electrodes on the left and right, any potential generated at the interfaces cancel each other (in principle), resulting in the system voltage being dependent only on the interaction of the glass membrane and the test solution. The potential difference between the two reference electrodes arises because of the exchange of Na^{+} ions by H^{+} ions on both sides of the glass membrane. H^{+} does not cross the glass membrane. The measuring part of the electrode, that is the glass bulb on the bottom, is coated both inside and outside with a ~10 nm layer of a hydrated gel. These two layers are separated by a layer of dry glass.

Thus, a glass electrode is essentially a galvanic cell that can be schematically represented as follows:

Reference electrode // Reference Solution // *Test Solution* // Reference electrode or more detailed with silver/silver chloride reference electrodes as follows:

$$\mathrm{Ag(s)/AgCl(s)/KCl(aq)//1\times 10^{-7}MH^{+}solution//glass\ membrane//}\textit{Test Solution}\mathrm{//}$$
$$\mathrm{junction//KCl(aq)/AgCl(s)/Ag(s)}$$

In this schematic representation of the galvanic cell, symmetry is observed between the left and the right membranes as seen from the center of the row occupied by the "Test Solution" (the solution whose pH must be measured).

The measured potential E is given as follows:

$$E = E^0 + RT/F \ \ln a(H^+) = E^0 - (RT/0.4343\,F) \cdot pH \tag{4.8}$$

Between measurements any glass and membrane electrode should be kept in a solution of its own ion. It is necessary to prevent the glass membrane from drying out because the performance is dependent on the existence of a hydrated layer, which forms slowly.

All glass pH electrodes have extremely high electric resistance ranging from 50 to 500 MΩ. Therefore, the glass electrode can be used only with a high input-impedance measuring device like a pH meter, or, more generally, an electrometer, that is, a high input-impedance voltmeter.

There are different types of pH glass electrodes; some of them have improved characteristics for working in an alkaline or acidic media. But almost all electrodes are sufficient for working in the most popular pH range from pH 2 to pH 12.

Finally, it should be mentioned that there are other types of exchange electrodes available. For these, many of the statements given for the glass electrodes are valid.

4.1.4 Potentiometry

Potentiometric titration is one of the most used electrochemical methods in analytical chemistry.

In "one-electrode" potentiometry the potential of an indicator electrode is measured with respect to a reference electrode.

According to the Nernst equation:

$$E_e = E^0 + RT/nF \ \Sigma \ \nu_i \ln a_i \tag{4.9}$$

One can follow the progress of a chemical reaction, if the activity a_i of a component in the solution influences the equivalence potential E_0 of the indicator electrode.

As an example an acid–base titration is used (Fig. 4.2). It shows the titration of 100 mL of a strong acid (c = 10^{-2} mol/L) with a strong base (c = 10^{-1} mol/L). The indicator electrode is a proton-sensitive electrode (glass electrode) and the reference electrode is a saturated calomel electrode.

The potentiometric titration can be used for analyzing redox reactions, precipitation reactions, complex formation reactions and so on.

Similar methods are amperometric titration, where the limiting current at the indicator electron is used for following a particular reaction, and the conductometric

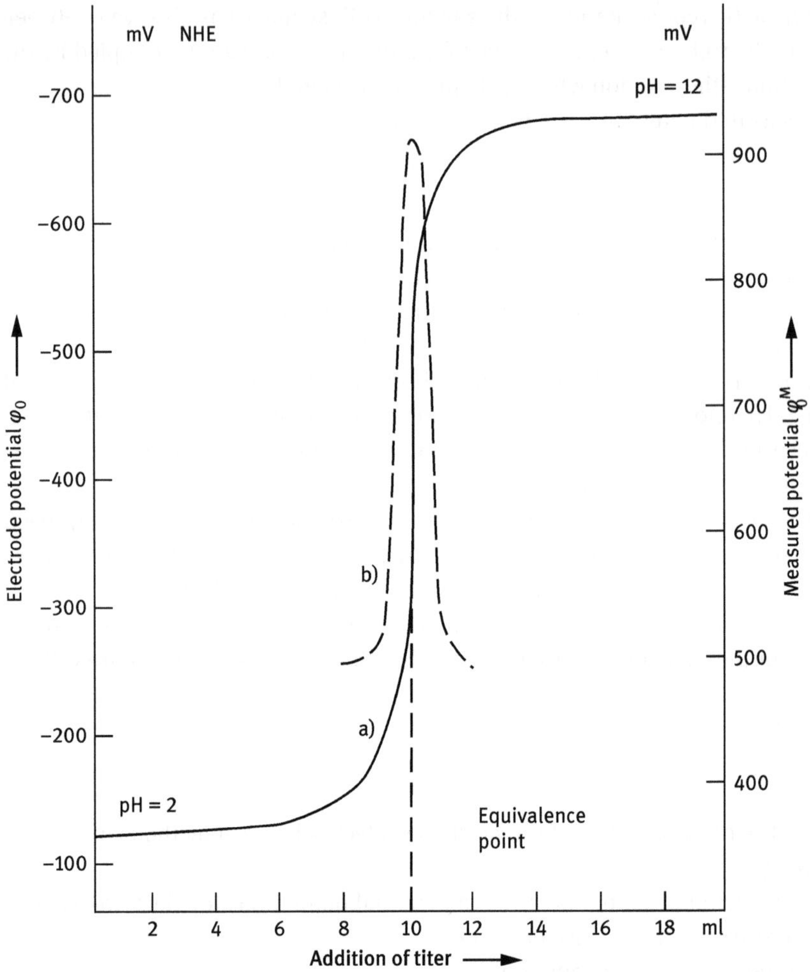

Fig. 4.2: Schematic presentation of a potentiometric acid–base titration. On the left ordinate the potential of the indicator electrode versus the normal hydrogen electrode is shown. On the right ordinate the potential of the indicator electrode versus a saturated calomel electrode is shown. The curve "a" represents the course of the titration curve, curve "b" is the first derivative of the titration curve close to the equivalence point (pH = 7).

titration where the conductivity of a solution is measured. More on this type of analytical application can be found in the literature.[52]

52 C.H.Hamann, W. Vielstich "Electrochemie" Wiley-VCH, Weinheim 2005, p. 561ff.

4.1.5 Coulometry

Coulometry utilizes the applied current or potential to completely convert an analyte from one oxidation state to another. In these experiments, the total current passed is measured directly or indirectly to determine the number of electrons passed. Knowing the number of electrons passed can indicate the concentration of the analyte or, when the concentration is known, the number of electrons transferred in the redox reaction can be determined. The common forms of coulometry include bulk electrolysis, also known as potentiostatic coulometry or controlled potential coulometry, as well as a variety of coulometric titrations.

4.1.6 Electrochemical sensors

Sensors are detecting elements delivering a signal, which is proportional to the concentration of a species of interest (analyte) or to a physical quantity (temperature, pressure, etc.) This signal is transformed by a transducer into an electric signal, which can be used for controlling the status of a system, more often it is used to start activities, such as closing or opening a valve, changing the temperature, adding a certain volume of a solution and so on and improving or stabilizing a system. Many of the sensors are ion-specific electrodes.

There exists a large number of principles used for the construction of sensors. In the following sections, we confine this report on sensors using electrochemical principles. One can distinguish potentiometric and amperometric sensors. The potentiometric sensor itself produces an electric voltage, which is directly measureable. These are often solid-state ionic conductors. At the amperometric sensor, a measurable current is produced. Other types of sensors are the resistive sensor where the conductivity is influenced by an analyte, preferably a gas. They are often established by inorganic metal oxide semiconductors (MOS), but also by conducting polymers or phthalocyanine. Capacitive sensors use the capacity of a condenser with gas-sensitive dielectric. They are often used for controlling the humidity of gases.

Gas sensors are important parts of safety systems. They are used for detecting gas leaks and can be interfaced with a control system. They can be used for detecting combustible, flammable and toxic gases or oxygen depletion.

Electrochemical gas sensors work by allowing gases to diffuse through a porous membrane to an electrode where it is either oxidized or reduced. Mostly a three-electrode system is used that consists of the working, counter and the reference electrodes. Such a miniaturized electrochemical cell may have a diameter of few millimeters. The concentration of the gas is proportional to the amount of current produced at the working electrode. All gas detectors must be calibrated on schedule. Since sensors are subject to corrosive elements or chemical contamination, a replacement may be required after about two years. Within this time it may they be recalibrated every three to six months.

Electrochemical gas detectors are used in a wide variety of environments such as raffineries, chemical plants, gas turbines, underground gas storage facilities and so on.

More details on electrochemical sensors can be found in literature.

4.2 Industrial electrochemistry

At the end of the nineteenth century after cheap electrical current was available by the discovery of the dynamoelectric principle, industrial electrochemistry received a boost from this invention. At the turn of the nineteenth to the twentieth century existed about 130 industrial plants for the electrolytic production of chlorine, chlorate, zinc, aluminum, copper and so on.

Compared to classical chemical processes the electrochemical processes may have advantages as well as disadvantages.

The yield of energy per kilogram of the product in electrochemical processes is often better than in chemical processes; however, heat energy is cheaper than electric energy. The electrolytic process needs not only the free enthalpy

$$\Delta G = n \cdot FE_c \tag{4.10}$$

of the reaction but in addition $n\,F\,\eta$ of the overvoltage η at the anode and the cathode and the loss because of the ohmic drop ΔU_Ω. Thus the real voltage at the electrolysis cell is given as follows:

$$U_c = E_c + \Sigma\eta + \Sigma\Delta U_\Omega \tag{4.11}$$

The voltage U_c of the cell is often 50–100% above the thermodynamic value.

An advantage of electrolysis is high selectivity of the reaction and purity of the product; the reaction occurs at normal temperature and pressure. Current and voltage can be easily controlled; therefore, automatic control of the process is easy. On the other side the financial investment for electrochemical plants are generally higher than for comparable chemical setups. Electrochemical plants need special electrodes for every process, special construction of the electrochemical cell and so on. In the following sections, some important and characteristic technical large-scale electrochemical processes will be described; however, in the frame of this booklet details cannot be described; we refer the interested reader to textbooks of electrochemistry.[53, 54]

---

53 C. H. Hamann, W. Vielstich "Electrochemistry", Wiley-VCH Weinheim 2005, p. 587ff.

54 Encyclopedia of Electrochemistry (ed. by Bard and Stratmann), Wiley-VCH 2003, Vol. 3.

Organic electrochemistry offers many reactions, which can be used in small-scale productions of interesting compounds. For the reader who is interested in this field, special textbooks are recommended.[55]

4.2.1 The electrochemical production of chlorine and sodium hydroxide

Chlorine is one of the most important fundamental substances for industrial chemistry. It is produced by electrolysis of aqueous sodium chloride solutions.

The cell reaction is as follows:

$$2\,NaCl + 2\,H_2O \longrightarrow Cl_2 + H_2 + 2\,NaOH \tag{4.12}$$

For the technical practicability of the electrolysis, one needs methods that avoid the transport of hydroxide ions from the cathode to the anode because this leads to the side reactions 4.13 and 4.14.

$$Cl_2 + 2\,OH^- \longrightarrow ClO^- + Cl^- + H_2O \tag{4.13}$$

and at the anode further oxidation of hypochlorite to chlorate according to the following reaction:

$$6\,ClO^- + 3\,H_2O \longrightarrow 2\,ClO_3{}^- + 4\,Cl^- + 6\,H^+ + 1.5\,O_2 + 6\,e^- \tag{4.14}$$

These side reactions produce impurities of hypochlorite and chlorate in sodium hydroxide and of oxygen in chlorine. Moreover, the current yield in respect to chlorine decreases considerably.

This is the reason why three different processes have been developed for this technical electrolysis.

Already in 1892 the "amalgam process" had been introduced. In this process for the cathode mercury is used. Sodium metal deposited at the cathode is dissolved in mercury forming a mercury/sodium alloy (amalgam). Thus this process hinders the formation of hydroxide ions and the problem arising from them. In a separate vessel, one gains hydroxide ions by washing the amalgam with water. The anode is made of titanium covered with ruthenium oxide.

The question appears: Why is at the cathode sodium deposited and not hydrogen? There is a high overvoltage of about 1.3 V for the hydrogen formation at mercury in neutral solution. Nevertheless, the deposition of sodium is expected at −2.71 V. Because of amalgamation sodium and mercury are deposited at −1.78 V NHE (0.2 % concentration of amalgam and 5M of sodium ions) and in the alkaline solution the

55 F. Beck,"Elektro-organische Chemie" Verlag Chemie, Weinheim 1974. Fr. Fichter,"Organische Elektrochemie" Th. Steinkopff, Leipzig 1942; (This book contains an enormous number of experimental results obtained since the beginning of organic electrochemistry).

potential of hydrogen formation is shifted to negative potentials by 0.64 V. Thus sodium amalgam formation occurs without the formation of hydrogen. At the anode chlorine is formed at about +1.7 V including an overvoltage of about 0.4 V. The decomposition voltage according to the reaction 4.15: is observed at 3.11 V:

$$2\,NaCl + 2x\,Hg \longrightarrow Cl_2 + 2\,NaHg_x \tag{4.15}$$

Because of mercury pollution in the environment, especially in water, the amalgam process has been more and more substituted by two other processes.

The alternative "diaphragm process" has already been developed in 1890. The migration of hydroxide ions to the anode is hindered by a plastic diaphragm stabilized by fibers of asbestos. In addition the separation by streaming of electrolyte in the direction of the cathode compensates the migration and diffusion of the sodium ion.

In the technical performance of the electrolysis, iron is used for the cathode and the anode is titanium covered with ruthenium oxide as an electrocatalyst. The volume of technical cells comes to about 15 m^3. For the electrolysis one uses solutions containing 310–320 g/L sodium chloride at a cell voltage of about 3.5 V and a temperature of 80–95 °C. After 24 hours the solution is concentrated by evaporation to 50% of its volume and contains 180–210 g/L sodium chloride and 120–140 g/L sodium hydroxide thereby precipitating sodium chloride. From the resulting lye (still containing about 1% sodium chloride) one obtains flakes of NaOH. At the anode evolves hydrogen and chlorine (about 3 t/day). The expended electrical energy is 2800 kWh/t, the theoretical value comes to only 1623 kWh/t.

The "membrane process" unifies the advantages of the diaphragm process (low cell voltage) and of the amalgam process (NaCl-free lye) by substituting the diaphragm by a membrane with high Na^+ conductivity. The membrane requires long-time stability in an aggressive medium. It was the development of perfluorosulfonate and perfluorocarboxylate membranes that enabled the technical use of this modification in the electrolysis. The main advantage of the membrane process is the lower cell voltage 3.2 V at 3–4 kA/m^2 in comparison to the diaphragm process. The salt water has to be free of ions like Ca^+, Mg^{++} and so on forming insoluble hydroxides in the membrane. The current yield comes to 95%; it is comparable to that of the diaphragm process. Because of the high technical standard the membrane process is preferred for new investments.

Finally, it should be mentioned that the electrolysis of aqueous hydrochloric acid is also important. It recovers chlorine from HCl resulting in many organic substitution reactions.

4.2.2 Production of metals

The electrochemical production of metals generally uses acidic aqueous solutions of metal sulfates. The reduction of the metal ions with a deposition potential negative to that of hydrogen needs cathodes with considerable hydrogen overvoltage. Numerous metals such as gold, silver, copper, lead, tin, nickel, cobalt, cadmium, chromium and zinc are obtained by electrolysis. The most important process is the production of zinc.

The cell for the zinc electrolysis consists of big concrete troughs (3 m, 1 m and the depth of 1.5 m) coated with lead or plastic. At the aluminum cathode, zinc is deposited from an aqueous solution containing about 240 g/L zinc sulfate and in addition 40 g/L free sulfuric acid. The cell voltage is 3–4 V, about twice the value of the decomposition voltage. At a current density of 1 kA/m^2 within 27 h, 32.7 kg zinc per m^2 is deposited; this corresponds to a layer of 5 mm thickness. The metal is removed mechanically from the cathode and pulled out of the electrolyte.

Electrochemical raffination is another industrial process. It is used for the purification of copper. In an aqueous electrolyte of copper sulfate, both the anode and the cathode are copper electrodes. Applying a small voltage (about 0.2 V) between the electrodes leads to dissolution of copper at the anode and deposition of copper at the cathode. The anode is "raw copper," containing impurities such as Fe, Ni, Co, Zn, As and "noble metals" such as Ag, Au, Pt. At a working potential of +0.5 V versus Normal Hydrogen Electrode (NHE), impurities with a normal potential more negative than the working potential of the anode are oxidized to the corresponding ions. The "noble metals" remain during the dissolution of the anode in the form of fine metallic particle sampling at the bottom below the anode to the so called anode sludge – a valuable byproduct. At a working potential of 0.3 V versus NHE only copper is deposited. This "electrolyte copper" is of very high purity.

4.2.3 Fused salt electrolysis

Many metals such as Li, Na, Be, Ca, Al and so on cannot be produced by electrolysis in aqueous medium. Their very negative normal potentials lead to hydrogen evolution at the cathode. This problem can be overcome by using organic solvents; however, this would be a very expensive solution of the problem.

The electrolysis of fused salts is a well-known method for circumventing the difficulties. Aluminum, magnesium and sodium are produced in large scale by this method. In a smaller scale, it is used for the preparation of Li, Be, B, Ti, Nb and rare earths.

The production of aluminum is the most important case. One uses aluminum oxide (Fp. 2,050 °C) in an eutectic mixture with cryolite (Na_3AlF_6). This mixture has a melting point below 1,000 °C. The decomposition potential of cryolite is higher than that of aluminum oxide. Thus from the electrochemical point of view the melt is an analog to a solution of aluminum oxide in cryolyte.

Figure 4.3 shows schematically the electrolytic cell. The cathode is liquid aluminum, the anode is "petrol coke," a special mixture of carbon and coal tar which is filled in iron vessels; it solidifies at high temperatures. Because of the formation of oxygen, the anode is burning down and must be readjusted during the electrolysis. Liquid aluminum is heavier than the eutectic melt; it can be discontinuously taken out.

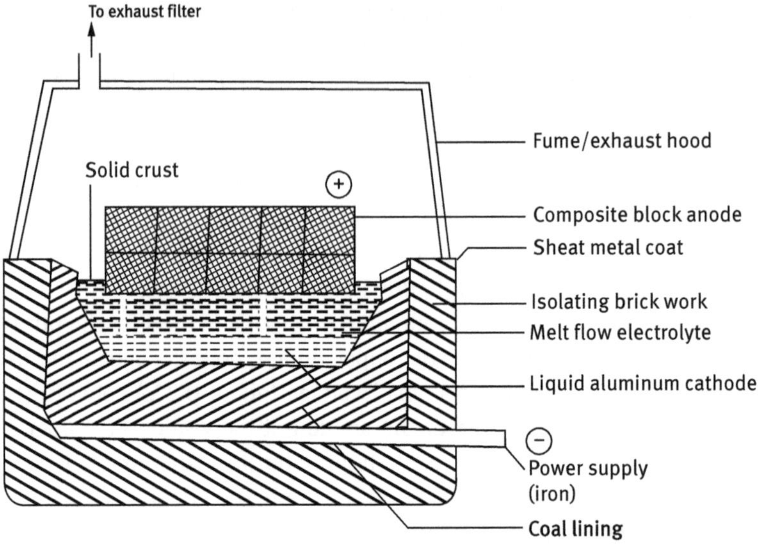

Fig. 4.3: Scheme of the melting flux electrolysis of aluminum. (The ground area of the "oven" is 4 x 8 m, the distance between anode and cathode is about 5 cm). The working temperature is about 1,000 °C; it is maintained by Joule heating.[56]

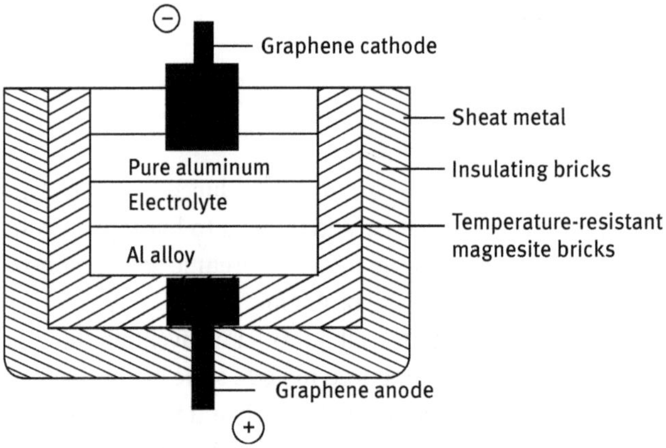

Fig. 4.4: Schematic view of a cell for aluminum raffination.

The cell voltage is about 4.2 V, the current density at the anode is between 6.5 and 8 A/m^2 and at the cathode 3–3.5 A/m^2. The production of one ton of aluminum needs 1900 kg aluminum oxide, 500 kg anode coal, about 40 kg cryolite and 12000–15000 kWh electrical energy, i.e. up to about 8 kWh per kg aluminium.

56 C. H. Hamann, W. Vielstich "Electrochemistry", Wiley-VCH Weinheim 2005, p. 453f.

Aluminum produced in this way contains about 0.1% silicon and 0.1% iron. It can be further purified using an electrochemical raffination. This supplies technical amounts of aluminum with a purity of 99.9%.

Recently, the "Alco–process" gained increasing significance. Aluminum is deposited from a melt of sodium chloride/lithium chloride in the ratio of 1:1 containing10–15% aluminum chloride. Graphite electrodes are used at a working temperature of about 700 °C. At the anode, chlorine is produced which is used to convert aluminum oxide into aluminum chloride.

The production of aluminum is a typical example for this kind of industrial electrochemistry. There are numerous other similar processes. Being in competition with chemical processes they are used if the latter are more expensive or are technically more complex than the electrochemical technique.

4.2.4 The electrolysis of water

The electrolysis of water delivers hydrogen at the cathode and oxygen at the anode. It is a popular idea to use it for the storage of superfluous electrical energy produced in wind and solar plants. However, applying industrial standards, the realization of this idea remains to be seen.

At present, only in close neighborhood with big water power plants, where very cheap electrical energy is available, the generation of hydrogen by electrolysis of water can compete with chemical processes.

The thermodynamic calculation of the depletion of water into hydrogen and oxygen predicts a depletion voltage of 1.23 V. However, there exists a considerable overvoltage of several 100th mV at the anode. Therefore, chloride-containing conducting salt has to be excluded. With sulfate-containing conducting salts, corrosion has been observed; therefore, electrolysis is performed in an alkaline medium.

The separation of the gases needs a suitable membrane between the electrodes (fig. 4.5). The electrodes are made of surface-manipulated nickel or steel nets. In order to minimize the resistance of the electrolyte the electrodes are placed close to the membrane. The gas bubbles have to be guided carefully into the current-free space. Gas bubbles between the electrodes may increase the cell resistance because of a decrease in the cross section. The single cells are combined to large packages producing up to several 100th Nm^3/h hydrogen gas. The classical electrolysis is performed at a cell voltage of 1.8–2.0 V, at current densities between 1–2.5 kA/m^2 and at a temperature of 80 °C. To the running system only chloride-free water can be added. The production of 1 Nm^3 ($p = 1$ atm, $T = 298$ K) hydrogen consumes 4 kWh electrical energy. For the transport and the storage of hydrogen, additional costs have to be taken into account.

Economical calculations demand an improvement of the process. This may be possible by reducing the overvoltage by detecting new electrode materials, by

The reactions a–g reveal the complexity of this industrial process. It is very sophisticated to guide such a process on a pathway that delivers high yields of the desired product.

A severe trap during the realization of the Monsanto process was the following reaction:

$$\underset{3}{CH_2-CH_2CN^{\cdot}} + e^- \longrightarrow \underset{7}{CH_2-CH_2CN^-} + H^+ \longrightarrow \underset{8}{CH_3CH_2CN} \tag{4.18}$$

The process runs in an aqueous medium where protons are readily available. Thus, the yield of adipic-dinitrile may be small because of the competition of reaction 4.16 with reaction 4.18 leading to propionitrile. A deeper analysis shows that the protonation reactions preferably occur near the surface. Using large unsolvated supporting ions turns out water molecules near the surface of the electrode; thus, supporting the dimerization reactions; protonation of dimers is carried out at a larger distance from the surface.

The distance between the stainless steel electrodes is only 2 mm; the cathode is covered with cadmium (high overpotential for hydrogen). The cell voltage is 4 V and the current density comes to 2 kA/m^2.

4.3 Electrochemical generation and storage of energy

4.3.1 Galvanic cells

4.3.1.1 Some general aspects

Galvanic cells are transportable sources for electric current. There exists a considerable number of different transportable sources for electricity depending on the special applications. In the following sections, we distinguish nonrechargeable cells ("primary cells"), generally called batteries and rechargeable cells ("secondary cells") that are also termed accumulators.

In galvanic cells, chemical energy is converted into electric energy. In the rechargeable cells the reverse process is possible; they can store electrical energy.

For constructing a galvanic cell, many possible combinations of anodes and cathodes are at our disposal. At the anode substances that can be electrochemically oxidized are used, for example, metals and hydrogen are used ("negative active mass") and at the cathodes electrochemically reducible substances ("positive active mass"), for example, metal oxides and oxygen are used.

For galvanic cells in practical use some requirements have to be fulfilled.

The minimum rest potential of a cell should be about 1 V and the working potential is at least 0.5 V. Small activation barriers (nearly reversible reactions) of the electrode processes are desirable, allowing large currents from the cell without breakdown of the cell voltage.

The active masses of a battery can discharge when no current is taken from the battery (self-discharging). Therefore the active masses must be chemically stable and protected by high overvoltage against such corrosion reactions. The relation between capacity and weight has to be taken into account. Ultimately, economical aspects play a role; this is the price of the active masses and the lifetime of a battery, that is, in case of rechargeable batteries the number of charging–discharging cycles.

The first usable "battery" was the Volta column. The column was established by using plates of copper and zinc, separated by felt, which was saturated by sulfuric acid. In fact this was a precursor of the Daniell-element. Stacking of such elements allowed rather high voltage and considerable capacity of a battery. At the early nineteenth century this was the only source of electricity. Of course, this "battery" was not very handy, but within a rather short time other batteries have been developed.

4.3.1.2 Nonrechargeable systems

Leclanché described in 1865 a battery consisting of a zinc rod, manganese dioxide and of ammonium chloride.

The reaction at the zinc anode is as follows:

$$Zn \longrightarrow Zn^{++} + 2\,e^- \tag{4.19}$$

The corresponding cathodic reaction is the reduction of $Mn^{4+} \rightarrow Mn^{3+}$:

$$2\,MnO_2 + 2\,H_2O + 2\,e^- \longrightarrow 2\,MnOOH + 2\,OH^- \tag{4.20}$$

Because of the formation of hydroxyl ions results ammonia in the electrolyte, which forms a complex with the zinc ions. This reaction in the electrolyte is:

$$Zn^{++} + 2\,NH_4Cl + 2\,OH^- \longrightarrow Zn(NH_3)_2Cl_2 + 2\,H_2O \tag{4.21}$$

The brutto cell reaction is:

$$2\,MnO_2 + Zn + 2\,NH_4Cl \longrightarrow 2\,MnOOH + Zn(NH_3)_2Cl_2 \tag{4.22}$$

The standard potential of the zinc electrode is −0.76 V and that of the manganese dioxide electrode is about +1.1 V. Both these potentials depend on the pH value; therefore, the theoretical value of 1.86 V is not reached; the practical voltage under open circuit conditions is between 1.5 and 1.6 V.

A lot of practical experience is involved in the production of these batteries. The rate of the anodic reaction at compact zinc is sufficient; however, for the cathode it is necessary to mix the manganese dioxide with carbon powder in order to gain the mass and conductivity necessary for its practical use.

In addition to the voltage the capacity of a battery is an important property. It is measured in Ah/kg (Amperehours per kilogram).One Amperehour is 3.6 10^3 J or 861cal, this is not too much energy. The capacity can be calculated from the thermodynamic data of the overall cell reaction. However, one observes considerable deviations from the thermodynamic values because of side reactions and failure of the electrode equilibria. The theoretical capacity of the Leclanché element is 155 Ah/kg at the terminal voltage ($i = 0$) 1.58 V. The real available specific capacity depends strongly on the conditions of discharging. For "dry batteries," reference values of the capacity are given under the condition of five hours lasting discharging capacity. Fast discharging decreases the capacity.

The Leclanch*é* battery has been further developed to the alkaline-manganese battery. It uses manganese dioxide and zinc but the electrolyte is KOH.

The anodic reaction is as follows:

$$Zn + 2\ OH^- \longrightarrow Zn(OH)_2 + 2\ e^- \tag{4.23a}$$

$$Zn(OH)_2 + OH^- \longrightarrow Zn(OH)_3{}^- \tag{4.23b}$$

The cathodic reaction is as follows:

$$MnO_2 + H_2O + e^- \longrightarrow MnO(OH) + OH^- \tag{4.24a}$$

$$MnO(OH) + H_2O + e^- \longrightarrow Mn(OH)_2 + OH^- \tag{4.24b}$$

A rechargeable version of this battery is feasible if only the first step of the manganese dioxide reduction is used.

4.3.1.3 The zinc–mercury oxide battery

This battery (often called "Mallory battery") has a high density of energy (110 Wh/kg) and long storage life (5–8 years). Small Mallory batteries are used in watches, cameras, calculators, ear trumpets and pacemakers.

The electrode reactions in an KOH electrolyte are as follows:

$$Zn + 2OH^- \longrightarrow Zn(OH)_2 + 2\ e^- \tag{4.23a}$$

$$Zn(OH)_2 + OH^- \longrightarrow Zn(OH)_3{}^- \tag{4.23b}$$

and at the positive mass containing about 90% mercury oxide and 10% graphite

$$HgO + H_2O + 2\ e^- \longrightarrow Hg + 2\ OH^- \tag{4.25}$$

This battery is, because of ecological aspects, in some retreat and may be substituted by lithium batteries.

4.3.1.4 Lithium batteries

There are several reasons for favoring lithium as a material for batteries: the voltage is E_0 = -3.045 V versus NHE, the theoretical capacity C_{th} = 3862 Ah/kg and the specific weight M = 6.939 g/ cm^3.

Using a suitable positive electrode, the cell voltage may be more than 4 V. Of course this needs appropriate nonaqueous solvents.

In the Li/SO_2 battery the electrolyte is a mixture of acetonitrile, propylene carbonate and sulfur dioxide (several bars) and 1.8 molar lithium bromide. The separator is a polypropylene folia together with folia of porous carbon. The system is enclosed in a gas-tight capsule. The overall cell reaction is as follows:

$$2\,Li + 2\,SO_2 \longrightarrow Li_2S_2O_4 \tag{4.26}$$

The equilibrium cell voltage is about 2.9 V, the theoretical capacity C_{th} comes to 1,095 Ah/kg.

Another system is the lithium/thionyl chloride battery. In this battery, thionylchloride is also the positive mass as the solvent. The supporting electrolyte is $LiAlCl_4$. These batteries have a cell voltage of 3.65 V and capacity is 1,470 Ah/kg.

The batteries have low self-discharge rates, which is equivalent to long working periods. On the other side, the batteries contain vey toxic and high-energetic material, which makes it necessary to absolutely prevent the damage.

There are other types of lithium-based batteries; for these, we refer the reader to special literature.[59]

4.3.1.5 Rechargeable batteries (accumulators)

Rechargeable batteries are of special interest in the framing of electrochemical energy storage. They offer transportable sources of electricity and can be used for longer time. The most interesting data of rechargeable batteries are cell voltage and storage capacity; other data are the number of charging–discharging cycles and the price of the stored current. In the following sections, some rechargeable batteries will be described.

The most common rechargeable battery is the lead accu. It has a long history; the first one has been build in 1860 by Planté. Since then this storage battery has been continuously improved. Modern lead accus have a voltage (i = 0) of 2 V and a capacity of 166 Ah/kg. This latter value is rather low because of the considerable weight of the accus (the Ah/volume is much more favorable). The electrode reactions have already been reported (p. 19) and shall not repeated here.

---

59 K.V.Kordesch “Batteries 1,” Marcel Dekker, New York 1974. J.P. Gabano (ed),“Lithium Batteries,” Academic Press, New York 1983.

Another rechargeable galvanic cell is the nickel–cadmium system introduced by Jungner in 1899. It uses cadmium and nickel oxide (NiOOH) in an alkaline solution. The electrode reactions are as follows:

$$Cd + 2\,OH^- \rightleftharpoons Cd(OH)_2 + 2\,e^- \tag{4.27}$$

and

$$2\,NiOOH + 2\,H_2O + 2\,e^- \rightleftharpoons 2\,Ni(OH)_2 + 2\,OH^- \tag{4.28}$$

Nickel–cadmium batteries are reliable and have better mechanical stability compared with the lead accumulator. They are preferably used in space and aviation projects because of their reliability.

The silver–zinc battery ("light-accumulator") utilizes a zinc electrode in an alkaline solution. The electrode reactions are as follows:

$$Zn + 2\,OH^- + 2\,e^- \rightleftharpoons Zn(OH)_2 \tag{4.29a}$$

$$Zn(OH)_2 + 2\,OH^- \rightleftharpoons Zn(OH)_3{}^- \tag{4.29b}$$

and

$$AgO + H_2O + 2\,e^- \rightleftharpoons Ag + 2\,OH^- \tag{4.30}$$

This battery has a considerable capacity; however, the high price and the relative short lifetime prohibit its broad distribution.

For further information on secondary batteries, the textbook of Vielstich and Hamann is recommended.[60]

4.3.2 Fuel cells

A fuel cell is an electrochemical cell that converts chemical energy from a fuel into electricity through an electrochemical reaction of hydrogen fuel with oxygen or another oxidizing agent.

The idea to supply an electrochemical element continuously with electroactive material goes back to the early nineteenth century. Grove was the first (1839, 1842) to construct a hydrogen–oxygen fuel cell. For the electrodes he used platinum in aqueous sulfuric acid. The first commercial use of fuel cells came more than a century later in NASA space programs to generate power for satellites and space capsules. Since then, fuel cells have been used in many other applications.

---

60 C. H. Hamann, W.- Vielstich, "Electrochemistry" Wiley-VCH 2005, p. 481ff.

Fuel cells are different from batteries in requiring a continuous source of fuel and oxygen (usually from air) to sustain the chemical reaction, whereas in a battery the chemical energy comes from chemicals already present in the battery. Fuel cells can produce electricity continuously for as long as fuel and oxygen are supplied.

Fuel cells are used for primary and backup power for commercial, industrial and residential buildings and in remote or inaccessible areas. They are also used to power fuel cell vehicles, including forklifts, automobiles, railways buses, boats and submarines.

Modern investigations on fuel cells started in context with astronautics. The hydrogen–oxygen fuel cell (Fig. 4.6) has been selected, which has a cell voltage of 1 V and a theoretical density of energy of about 1,000 Wh/kg. The active material (H_2, O_2) reacts on special catalyst electrodes in acidic or alkaline aqueous electrolyte.

The electrode reactions are in alkaline medium:

$$H_2 + 2OH^- \longrightarrow H_2O + 2\,e^- \tag{4.31a}$$

$$1/2\ O_2 + H_2O + 2\,e^- \longrightarrow 2\ OH^- \tag{4.31b}$$

and in acidic solution:

$$H_2 \longrightarrow 2\,H^+ + 2\,e^- \tag{4.32a}$$

$$1/2\ O_2 + 2\,H^+ + 2\,e^- \longrightarrow H_2O \tag{4.32b}$$

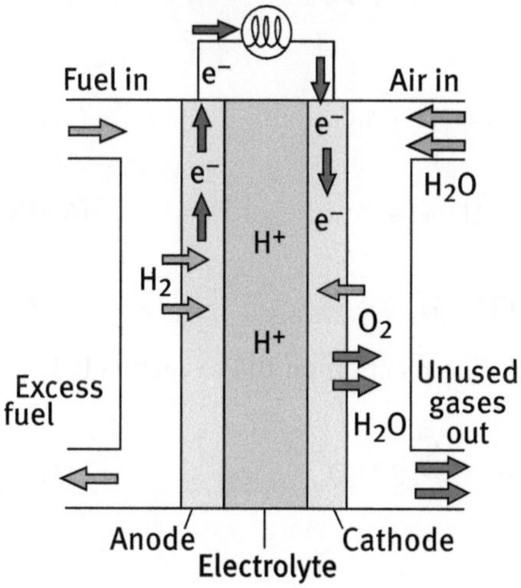

Fig. 4.6: Scheme of the hydrogen/oxygen fuel cell. (Source: I. Sakurambo)

The concentration of hydrogen and oxygen in aqueous solution is only 10^{-3} mol/L at room temperature. This low value and rather long way of diffusion results in low reaction rates. The development of the hydrogen–oxygen fuel cell has been described in some detail by C.H. Haman and W. Vielstich.[61]

It is interesting to get some data of the famous "Apollo cell." The cell contained stacks of H_2–O_2 fuel cells. The normal output of an "Apollo unit" was 1.12 kW at 28 V. The total weight of three units was inclusive weight of the tank, that is, 810 kg; during a flight of 10 days up to 500 kWh electrical energy was available. For comparison: For the same amount of electrical energy 10–12 t lead accumulators or 4 t light accumulators (Ag/Zn) would have been necessary.

The electromobility requires sources for electricity, which combine high capacity with huge power. High capacity is available if liquid fuel from a fuel tank can be used in the fuel cells. In the 1970s, military research developed a low-temperature liquid fuel cell using hydrazine as fuel and KOH as the electrolyte. The overall cell reaction is as follows:

$$N_2H_4 + 4\,OH^- \longrightarrow N_2 + 4\,H_2O + 4\,e^- \quad (4.33)$$

The theoretical energy density of this cell is 3,850 Wh/kg, the running system delivered up to 500 Wh/kg. Hydrazine is very toxic; therefore, there was a demand for other liquid fuels. Methanol, glycol and formaldehyde have been used in fuel cells.

We describe the direct methanol fuel cell (DMFC, fig. 4.7). In this fuel cell in acidic electrolyte two different pathways for the oxidation of methanol have been observed. They are the following reactions:

(Pathway I)

$$CH_3OH + H_2O \longrightarrow CO_2 + 6\,H^+ + 6\,e^- \quad (4.34)$$

(Pathway II)

$$CH_3OH \longrightarrow CO_{ad} + 4\,H^+ + 4e \quad (4.35a)$$

$$H_2O \longrightarrow OH_{ad} + H^+ + e^- \quad (4.35b)$$

$$CO_{ad} + OH_{ad} \longrightarrow CO_2 + H^+ + e^- \quad (4.35c)$$

Pathway II is more efficient than pathway I. The reaction on the oxygen side is as follows:

$$1/2\,O_2 + 6\,H^+ + 6\,e^- \longrightarrow 3\,H_2O \quad (4.36)$$

61 C.H. Hamann, W. Vielstich "Elektrochemie" 4.Aufl. Wiley-VCH, Weinheim 2005, p. 533ff.

Fig. 4.7: Demonstration model of a direct-methanol fuel cell. The actual fuel cell stack is the layered cube shape in the center of the image (Source: NASA).

From these reaction results the overall cell reaction is as follows:

$$CH_3OH + 3/2\,O_2 \longrightarrow CO_2 + 3\,H_2O \qquad (4.37)$$

Unfortunately, there is a severe shortcoming. The adsorption of the intermediate CO blocks the catalyst; this problem could be solved by using CO tolerant metal combinations such as Pt/Ru alloys. The theoretical equilibrium voltage is 1.21 V; using the cell 60–80% of this value is realistic. This fuel cell is of special interest for electromobility.

Typical values are 200 mA at 60 °C working temperature, 1 kWh per 1 L of methanol and 1mg Pt/Ru alloy per cm^2 area of the electrodes. These data correlate with 20% of the total efficiency. Some reliable DMFC systems are commercially available; however, it seems there is still a long distance needed to reach the road to success.

Of course, there is at present a considerable research activity in this field that creates improved DMFCs, a description of these activities is however beyond the scope of this book. The interested reader is referred to the literature.[62]

In addition to the proton exchange membrane fuel cells (PEMFC) there are other types of fuel cells such as phosphoric acid fuel cell (PAFC), solid acid fuel cell (SAFC), alkaline fuel cell (AFC) and high-temperature fuel cells. They all consist of an anode, a cathode and an electrolyte. Fuel cells may be classified by the type of electrolyte used and by the difference in start-up time ranging from 1 s for PEM fuel cells or PEMFC to 10 min for solid oxide fuel cells. Here we will make only some comments on high-temperature fuel cells. These are solid oxide fuel cells (SOFCs); they use a solid material, which is most commonly a ceramic material called yttria-stabilized zirconia as the electrolyte. Because SOFCs are entirely made of solid materials, they are not limited to the flat plane configuration of other types of fuel cells and are often designed as rolled tubes. They require high operating temperatures (800–1000 °C) and can be run on a variety of fuels including natural gas.

In addition to electricity, fuel cells produce water, heat, depending on the fuel source, very small amounts of nitrogen dioxide and other emissions. The energy efficiency of a fuel cell is generally between 40–60%; however, when the fuel cell's waste heat is used for heating a building in a cogeneration system, this efficiency can increase to 85%. This is significantly more efficient than traditional coal power plants, which are only about one-third energy efficient.

A compilation of different types of fuel cells is given in Table 4.1.

4.3.3 Supercapacitors

A supercapacitor (SC) is a high-capacity capacitor with capacitance values much higher than other capacitors (but lower voltage limits) that bridge the gap between electrolytic capacitors and rechargeable batteries. They typically store 10 to 100 times more energy per unit volume or mass than electrolytic capacitors. SCs can accept and deliver charge much faster than batteries, and tolerate many more charge and discharge cycles than rechargeable batteries.

SCs are used in applications requiring many rapid charge–discharge cycles rather than long-term compact energy storage within cars, buses, trains, cranes and elevators, where they are used for regenerative braking, short-term energy storage or burst-mode power delivery. Smaller units are used as memory backup for static random-access memory (SRAM).

62 C.H.Hamann, W.Vielstich, Wiley-VCH, Weinheim 2005 p. 532ff and literature cited there.

Table 4.1: Properties of hydrogen–oxygen fuel cells.

Type	Temperature	Electrolyte	Electrodes	Application	% yield	Remarks*
Alkaline fuel cell (AFC)	90 °C	KOH (30%)	Carbon noble metal	Military space	About 50%	–
Solid Polymer (SPFC)	≤ 80°	Polymers H^+ conductivity	Thin noble metal layers	Electricity, heat supply*	About 55%	–
Phosphoric acid (PAFC)	≈ 220°	Conc. H_3PO_4 in matrix	Graphite + noble metal	Small power stations	40–55%	–
Molten carbonate (MCFC)	650°	Molten Li_2–K_2CO_3**	Ni, Ni-oxide with Li	Power stations	55–65%	–
Solid Oxide (SOFC)***	1,000°	O-conducting $ZrO_2 \times Y_2O_3$	Ni, $LaMnO_3$ doped	Power stations electromobiles	55–65%	

Remarks: * The first SPFC-submarine of the German navy, *U31*, started on 20.3.2002
** Molten carbonates of lithium and potassium in a ceramic matrix
*** This system is still in experimental development

Unlike ordinary capacitors, SCs do not use the conventional solid dielectric, but rather, they use electrostatic double-layer capacitance and electrochemical pseudocapacitance, both of which contribute to the total capacitance of the capacitor (fig. 4.8).

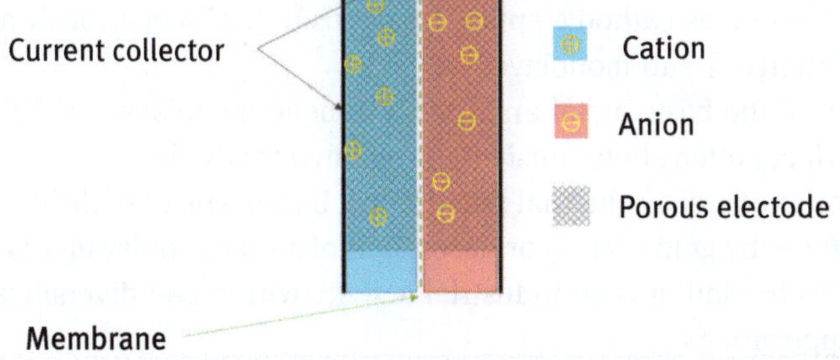

Fig. 4.8: Schematic illustration of a supercapacitor (Source: Z. Qi, G.M. Koenig, https://doi.org/10.1116/1.4983210).

4.4 Metal deposition and corrosion

4.4.1 Metal deposition

Metal deposition is important for improving the properties of metal surfaces and for protecting metal surfaces against corrosion. The deposition of metals from aqueous or nonaqueous solutions follows the electrode process described for the metal–ion electrode, which can be described by the following reaction:

$$\mathrm{Me(lattice)} + \mathrm{m\ X} \rightleftharpoons (\mathrm{Me}^{z+}\mathrm{X_m})(\mathrm{solution}) + \mathrm{z\ e^-} \tag{4.38}$$

X symbolizes complex forming species such as H_2O and NH_3 or CN^-, Cl^-. The surface of a solid metal electrode is partially a single crystal surface. The deposition of metal atoms on such a surface indicates the growth of the crystal. This growth results from the formation of new layers upon the older ones. This occurs by addition of metal in monoatomic steps. Two principal possibilities appear for this process. The metal may be discharged in the outer Helmholtz plane of the double layer and a so-called adatom is formed. The adatom moves toward the surface of the electrode and finally finds a binding site preferably at a step. The second possibility is the following: The adatom is bound at the outermost plane of the single crystal and forms with additional adatoms a critical two-dimensional germ. Another possibility is that additional adatoms lead to the enlargement of a step, which is generated by screw dislocation that winds up to a nondisappearing spiral step.

Surface diffusion can be the rate determining step for metal deposition. (For details, see.[63] If this is the case one observes a crystallization overpotential.)

The deposition of a metal atom up to one monolayer according to the following reaction

$$\mathrm{Me}^{z+} + \mathrm{z\ e^-} \longrightarrow \mathrm{Me}^0 \tag{4.39}$$

often occurs at more positive potentials than the thermodynamic potential required by the Nernst equation. This effect is called under-potential deposition (UPD). The effect of UPD can be observed in cyclic voltammograms if very slow scan rates (1–20 mV/s) are used. One observes cathodic spikes by deposition of metal ions and anodic spikes by dissolution of a sub-monolayer (Fig. 4.9).

At a suitable choice of the basic metal and the UPD metal-monolayers of UPD metals the electrode surfaces often show considerable catalytic activities.

The metal deposition is of great industrial importance. Improvement of the electric conductivity of surfaces by gold plating or decoration of surfaces of jewelry is a well-known example. Electroplating is an industrial activity with broad diversifications of methods and applications.

63 C.H. Hamann, W. Vielstich "Electrochemistry" p. 225f.

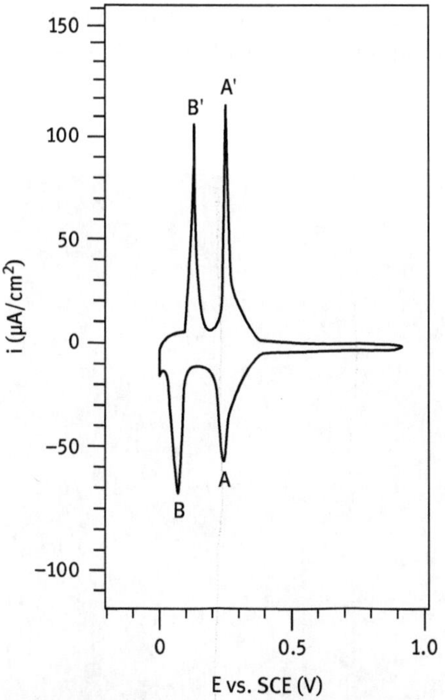

Fig. 4.9: Under potential deposition of Cu on Au <111>. Peaks A and B refer to the two step reduction of Cu++, A' and B'reveal the reoxidation of the Cu layers. Notice the low current density.

4.4.2 Corrosion

A piece of metal is stable for infinite time, provided it is stored under ultrahigh vacuum; if the metals are not isolated in this way, most of them become instable. The destruction of a metal beginning at its surface is called corrosion.

Corrosion is an electrochemical process and has to be distinguished from the term degradation, which means destroying of material by heat, light and so on by chemical reactions.

The anodic dissolution of a metal is observed in the Daniell-element, where we have the zinc anode dissolved by producing zinc ions. The copper cathode takes the electrons via the external circuit and produces copper metal from copper ions. The principle of corrosion is the same. The corroding metal is the anode producing metal ions; however, the corresponding cathodic reaction is either the reduction of protons to hydrogen or oxygen to O^{2-} (resulting in H_2O or OH^-). The second difference to the galvanic Danielle-element is that anode and cathode are not separated by a diaphragm but we have a short-circuit energy-producing cell.

As an example, the corrosion of iron will be described.

Everyone knows rust; this brownish crust on pieces of iron. Corrosion of iron is the process from which rust results. Iron corrodes readily in carbon dioxide containing water or humid air (fig. 4.10) .

Fig. 4.10: Rust, the popular corrosion product of iron (Source: Roger McLassus[64]).

The formation of rust is an electrochemical process; it can be formulated as a two-step process according to the following reactions:

$$Fe + 2\ H_2CO_3 \longrightarrow Fe^{++} + 2\,HCO_3{}^- + H_2 \tag{4.36}$$

$$2\,Fe^{++} + 4\,OH^- + 1/2\,O_2 \longrightarrow 2\,FeO(OH) + H_2O \tag{4.37}$$

Iron is a single redox electrode. The surface of the electrode electrons reduce protons in the first step followed by the reduction of oxygen in the second step.

Warning of corrosion can be stated simply on the basis of equilibrium thermodynamics.

64 https://commons.wikimedia.org/wiki/File:Rust_and_dirt.jpg

The standard equilibrium potential of the pair Fe/Fe^{2+} is −0.44 V. The half reaction $Fe^{2+} + 2e^- \longrightarrow Fe$ is coupled to another half reaction with a standard redox potential more positive than −0.44 V; the standard potential of the redox pair H^+–H_2 is $E^0 = 0$ V and that for the pair O_2/OH^- is $E^0 = 0.40$ V. From the difference between the potentials of the anodic and the cathodic half cell according to $\Delta G = -n\,F\,\Delta E$ the free enthalpy ΔG of the electrode reactions can be calculated. If ΔG is negative a corrosion process occurs.

In Table 4.2 the overall energetics of some spontaneous corrosion reactions are compiled.

Table 4.2: Overall energetics of some spontaneous corrosion reactions.

Metal	Hydrogen type oxygen type $pH_2 = 1$ at $pO_2 = 0.21$ at 298 K	
	Solid product	Free energy change at 25 °C Cal g-mol^{-1} metal
Silver	Ag_2O	+27000–1080
Copper	CuO	+24800–37450
	$Cu(OH)_2$	+27800–28300
	Cu_2O	+9500–18600
Lead	PbO*	+11500–44600
Nickel	$Ni(OH)_2$	+7800–48500
Cadmium	$Cd(OH)_2$	+600–55600
Iron	$Fe(OH)_3$	+4700–80000

Of course, the thermodynamic discussion of corrosion reveals only the possibility that the exposure of a metal to a particular environment leads to its decay. From the practical point of view of great interest is the rate of corrosion. We need to know how rapidly a thermodynamically unstable system will corrode and whether it is possible to decrease the corrosion rate.

The corrosion rate may depend on different kinds of overpotentials.

On the other side, the primary product of corrosion may form a layer on the surface protecting against further corrosion. The most prominent example is the oxide layer on aluminum.

4.4.3 Electrochemical etching

Electrochemical etching is a very important technique preferably applied for the fabrication of microstructures. The fabrication of microstructures follows several steps. The first step is the generation of a mask. This mask is produced by projecting the required structure on the layer of a photo resin. The exposed parts of the resin layer may be dissolved and thus the underlying material, that is, silicon is open for etching. For etching, silicon acidic solutions (HF) or alkaline solutions (KOH) may be used. This kind of etching is a kind of controlled corrosion. Etching of silicon is used in the fabrication of chips and microdevices such as vibration detectors. The spatial resolution depends on the spatial structure of the mask and the imaging optics. The present limits are 500 nanometers and less.

Another kind of electrochemical etching is the etching process that involves the use of a solution of an electrolyte, an anode and a cathode. The metal to be etched is connected with the positive pole (anode) of a source of direct electric current, the cathode should be of the same metal and the electrolyte should contain the same metal ions. As soon as current flows the metal at the anode is dissolved and finally deposited at the cathode. If the metal carries a structured mask, the structure of the mask is engraved in the metal. This is often used for marking metallic objects.

In a highly improved version of electrochemical etching, an STM electrode is used as one of the electrodes. If this electrode on the metallic object is positioned by a piezoelectric drive, structures with nanometer structure can be produced.

Finally, electropolishing should be mentioned. This is a commonly used technique for metal finishing in industries. For instance, the stainless steel drums of washing machines are electropolished. In addition, ultrahigh vacuum components are typically electropolished to have a smoother surface for improved vacuum pressure outgassing rates. Electropolishing is also commonly used for preparing thin metal samples for transmission electron microscopy and so on because the process does not cause mechanical deformation of surface layers observed with mechanical polishing.

Appendices

Appendix A

Basic principles of thermodynamics

This text is a very shortcut outline of concepts of thermodynamics for readers who are not at all familiar with thermodynamics. It is intended to bring the reader into the position to understand the basic principles of thermodynamics of electrochemistry.

Thermodynamics deals with the macroscopic state of the matter and with changes of this state. This state is characterized by independent variables such as temperature, pressure and volume. The microscopic structure and time are not considered in thermodynamics.

System and environment

We define an arbitrary object enclosed by walls with defined properties. This is the thermodynamic system to which refer all our considerations. Outside the system is environment. According to the properties of the wall we distinguish between different systems. If the exchange of energy and matter between the system and the environment is possible, then the system is an "open system." A "closed system" can exchange energy with the surrounding, but not matter. If neither energy nor matter can be exchanged with the environment, then we have an "isolated system." The electrochemical cells are generally closed systems.

Another important term is "phase." A phase is a part of a system where no differences of physical properties appear. For instance, in electrochemistry a metal electrode or the solution of an electrolyte is a phase. At the phase boundary, for example, metal/electrolyte, the electrochemical potential changes.

The simplest system suitable to introduce a newcomer into thermodynamics is a gaseous system. How the volume of a gas depends on temperature and pressure? How can the state of a gas be described?

Force, pressure and energy

The concepts of force, pressure and energy occur throughout chemistry and of course electrochemistry. It is important to be familiar with their units,

The basic unit of force is Newton (N): 1 N = 1 kg m/s^2, which may be illustrated. To accelerate a mass of 1 kg by 1 m/s^2 you need the force 1 N. Accelerating in this example means: starting with the mass on rest after 1 s it shall have the velocity

https://doi.org/10.1515/9783110437393-006

1 m/s, after 2 s the velocity is 2 m/s and so on. The acceleration due to force of gravity is 9.8 m/s^2. A small apple (≈0.1 kg) on a tree is subject to a force of gravitation of about 1 Newton.

Pressure is given by the force per area. The basic unit is called "Pascal" (Pa): 1 Pa = 1 N/m^2. An often used unit for pressure is the "atmosphere" (atm), and 1 atm is exactly 101,325 N/m^2, that is, ≈10^5 N/m^2.

The basic unit of energy is 1 Joule (J), 1 J = 1 N m. This may be illustrated. Raising a book of about 1.5 kg needs about 15 J energy. Each pulse of the human heart consumes about 1 J of energy. When 1 kJ of energy is dissipated in the form of heat in 50 mL water the temperature rises by about 5° C, which means, to make a small cup of tea requires about 150 kJ of energy.

Other units of energy are also used. The popular one is the calorie (1 cal = 4.18 J), which is a unit for energy in the form of heat. One calorie is necessary to rise the temperature of 1 mL of water by 1 °C.

Equation of state

How can we characterize a thermodynamic system? This shall be illustrated at the simplest system available. It is a volume of gas in a cylinder with a movable punch. Our system has a defined volume V, a defined temperature T and a defined pressure p. The pressure p of the gas is given by the weight on the punch. The gas counteracts the force, which the weight due to gravitation puts on the gas.

In principle, we can characterize the state of the gas by volume, pressure or temperature

The very early experiments of Boyle and Mariotte and of Gay-Lussac revealed that the volume, the pressure and the temperature for a given amount of gas (n moles) are related by the equation

$$p \cdot V = n \cdot R \cdot T \tag{A1}$$

n is the number of moles of gas under consideration (1 mol gas = 22.4 L at T = 298 K and p = 1 atm), R is the "gas constant." It is a universal constant independent from the nature of the gas (R = 8.314 J/mol/K = 1.98 cal/mol/K). A gas that follows eq. (A1) is called "perfect" or "ideal gas." Equation (A1) is the equation of state of an "ideal" gas.

At a given constant temperature T the product $p \cdot V$ is constant. For a deeper understanding of this equation, we analyze the dimension of the product $p \cdot V$; from p (N/m^2) · V (m^3) results the dimension N · m. This is the dimension of energy; therefore, the product p V gives the energy of the gas. It is immediately seen that this energy depends on the temperature T. (We mention that the temperature of rare gases – only spherical atoms – depends on the average velocity of the atoms. Heating this gas means enhancing the average velocities of the gas molecules.)

In principle, we can choose which of p, V, T shall be taken as the variables. Of course for describing a gas the volume is the clearest entity for describing the state of the gas. We see that the volume is a function of two variables:

$$V = f(p, T) \tag{A2}$$

We see from eq. (A1) that the response of the volume to changes in pressure at a constant temperature is given by

$$(V)_T = nRT \cdot 1/p \tag{A3}$$

At constant temperature, volume is inversely proportional to pressure. This relation has already been detected by Boyle in 1664.

The response to changes of the temperature at constant pressure has been formulated by Gay–Lussac in 1802 after first experiments from Charles (1787):

$$(V)_p = nRT \tag{A4}$$

The volume at constant pressure depends linearly from the temperature.

How can we describe the changes of the gas volume when both pressure and temperature change simultaneously?

We introduce the term "total differential" of V:

$$dV = \partial V/\partial p \, dp + \partial V/\partial T \, dT \tag{A5}$$

We illustrate the meaning of the total differential with a simple example: the area F of a rectangular area is given by $F = x \cdot y$. Changing this x and y by a infinitesimal amount dx and dy changes the area by dF:

$$F + dF = (x + dx)(y + dy) = xy + y\,dx + x\,dy + dx\,dy$$

$dxdy$ is negligibly small and can be neglected. We get dF by the sum of two partial changes

$$dF = y\,dx + x\,dy = \partial F/\partial x \, dx + \partial F/\partial y \, dy \tag{A6}$$

with $y = (\partial F/\partial x)_y$ and $x = (\partial F/\partial x)_x$. The total change of F is the sum of two partial changes. Total differentials and their derivations play a dominant role in classical thermodynamics.

Energy, heat and work

We come back to the system consisting of a cylinder with a movable punch and filled with a gas. We have an open system. The walls of the cylinder are diathermic, that is, they can be crossed by energy in form of heat. In addition, we have to make an agreement concerning the amount of material in our system. In thermodynamics, it is customary to use molar amounts if not otherwise stated. We refer to 1 mol or n mol of material. One Mol gas contains $6.02 \cdot 10^{23}$ atoms or molecules.

What happens when we transfer heat from the environment to gas? The gas will simultaneously expand and increase its temperature.

By expansion of the gas, the moveable punch is lifted upward, which means there is an output of work from the system. Work is done if a process can be used to bring about a change in the height of a weight somewhere in the environment. In addition, we observe increasing temperature of the system. This means energy is stored in the system. In our simple example, the system is a volume of a monoatomic gas where the average velocity of the atoms is enhanced. This energy is called the internal energy U of the system. In this specific system, the internal energy is given by the sum of kinetic energies of the atoms.

A change ΔU of the internal energy of the closed system is described by the difference of U_i, the internal energy of the initial state of the system, and U_f, the internal energy of the final state of the system

$$U_i - U_f = \Delta U = Q - W \tag{A7}$$

Q is the heat which is transferred from the environment into the system, W is the work, which the system can transfer to the environment. In thermodynamics, the amount of change in U is generally expressed by dU:

$$dU = dQ + dW \tag{A8}$$

The work W is given by $W = p \cdot V$ at constant temperature and therefore

$$dW = p\,dV + Vdp \tag{A9}$$

At constant pressure, $dp = 0$, U is a function of the temperature and the volume of the system, and the change of U can be expressed by the total differential of U:

$$U = f(T, V) \tag{A10}$$

and

$$dU = (\partial U/\partial T)\,dT + (\partial U/\partial V)dV \tag{A11}$$

Equations (A10) and (A11) show the form how changes of states containing only a single component are described generally in thermodynamics.

Before we continue with thermodynamics we have to be informed on the sign convention in thermodynamics.

The sign convention

We use the convention that dQ denotes the heat supplied to the system and dW denotes the work done by the system.

When dQ is positive, it signifies that heat is transferred into the system and that this has contributed to an increase in the internal energy.

When dQ is negative, it signifies that heat has flown out of the system and has contributed to a decrease in the internal energy.

If the system gains an amount of heat dQ, the environment loses the same amount of heat.

When dW is positive it signifies that work (mechanical energy) has been transferred to the system and thus has contributed to an increase in the internal energy.

On the other side, when dW is negative it signifies that work has been transferred to the environment. This leads to a decrease in the internal energy.

The origin of thermodynamics goes back to the problem of optimizing steam machines where you put in heat and you expect high yields of mechanical energy coming out. In a battery we put in chemical energy and get out electrical energy.

Now we can continue thermodynamic considerations.

Provided no other kinds of energy (e.g., electrical energy) are exchanged with the surrounding and the pressure in the surrounding is exactly equal to the pressure of the gas itself ($dp = 0$), eq. (A8) can be written as

$$dU = \delta Q_{\text{rev}} - p\,dV \tag{A12}$$

δQ_{rev} means exchange of heat in a reversible change of state, $dp = 0$.

The partial differential quotients in eq. (A11) have a physical meaning:

$$(\partial U/\partial T)_V = c_V \tag{A13}$$

is the molar heat at constant volume, that is, the energy which is necessary to increase the temperature of 1 mol of material by $dT = 1°$ C. If it refers to 1 g of material it is called specific heat: the specific heat of water at 25° C is 1 cal.

Systems with more components

In general, a system contains not only one component, but more components $n_1, \ldots, n_k$. Therefore, we have to extend eq. (A10) into

$$U = f(T, V, n_1, \ldots, n_k) \tag{A14}$$

And the total differential of the internal energy U then comes to

$$dU = (\delta U/\delta T)_{p,n} dT + (\delta U/\delta V)_{T,n} dV + (\delta U/\delta n_1)_{T,V,n\neq n_1} dn_1 + \cdots + (\delta U/\delta n_k)_{T,V,n\neq n_k} dn_k \quad \text{(A15)}$$

Enthalpy H

Now we will have a look at two different changes of state in a closed system ($dn = 0$) – the isochoric and the isobaric changes of state.

For the isochoric change of state holds $dV = 0$ and consequently

$$(dU)_V = \partial Q_{rev} \quad \text{(A16)}$$

During the transition from state J to state II in an isochoric process in a closed system, the change of internal energy is equal to the amount of heat exchanged with the environment.

In an isobaric change of state because of $dp = 0$ holds

$$dU = \partial Q_p - p\,dV \quad \text{(A17)}$$

and if we look again on the change from the state I to state II we get

$$U_{II} - U_I = Q_p - p\,(V_{II} - V_I) \quad \text{(A18)}$$

or

$$(U_{II} + p\,V_{II}) - (U_I + p\,V_I) = Q_p \quad \text{(A19)}$$

The heat Q_p which is exchanged with the environment in an isobaric process is not the change ΔU of the internal energy but of a new function of state called the enthalpy H:

$$H = U + p\,V \quad \text{(A20)}$$

$$(H_I - H_{II}) = Q_p \quad \text{(A21)}$$

$$H = f(T, p, n_1, \ldots, n_k) \quad \text{(A22)}$$

$$dH = (\delta H/\delta T)_{p,n} dT + (\delta U/\delta p)_{T,n} dp + (\delta H/\delta n_1)_{T,p,n\neq n_1} dn_1 + \cdots + (\delta H/\delta n_k)_{T,p,n\neq n_k} dn_k \quad \text{(A23)}$$

The molar heat Q_p exchanged with the environment at constant pressure is the difference ΔH of the enthalpy (A21); the heat capacity C_p at constant pressure is the partial derivative of H versus T:

$$C_p = (\partial H/\partial T)_p = (\partial Q/\partial T)_p \quad \text{(A24)}$$

Relations (A20)–(A24) refer to n mol of a pure homogeneous material. Division by the number of moles n yields the molar internal energy u and the molar enthalpy h:

$$u = U/n \text{ and } h = H/n \tag{A25}$$

The molar heat capacities are

$$c_V = (\partial u/\partial T)_v = C_V/n \tag{A26}$$

$$c_p = (\partial h/\partial T)_p = C_p/n \tag{A27}$$

At this point, we make a break in our presentation of basic thermodynamics. Readers who are interested in a more extended and deeper discussion are referred to textbooks of physical chemistry.[64] We continue our discussion with some other aspects that are relevant for electrochemistry.

As pointed out, the electrochemical cell is generally a closed system from the thermodynamic point of view. No materials is exchanged with the environment; however, the composition of the materials in the cell can change by a chemical reaction, for example, the overall cell reaction. A chemical reaction is connected with energetic changes in the system.

In a closed system with chemical reactions, changes in the amount of a component are not arbitrary but connected with each other by a given equation of reaction. A typical example is a chemical reaction

$$|\nu_A|\,A + |\nu_B|\,B \rightarrow |\nu_C|\,C + |\nu_D|\,D \tag{A28}$$

where ν_i are the stoichiometric factors. The stoichiometric factors of the educts (A, B) are taken to be negative and those of the products (C, D) are positive.

The internal energy and the enthalpy can be expressed by T, V, p and the amount n_i of different components of i:

$$U = f(T, V, n_1, \ldots, n_i) \tag{A29}$$

$$H = f(T, p, n_1, \ldots, n_i) \tag{A30}$$

$$dn_A/\nu_A = dn_B/\nu_B = dn_C/\nu_C = dn_D/\nu_D \tag{A31}$$

It is more simple instead of using the differential turnovers of each component for the thermodynamic description of the reaction to define the new quantity ζ which allows a clear description of the progress of a reaction. It is called turnover number and indicates the number of reaction turnovers. It seems to be more clear to call it formula conversion number because for $d\zeta = 1$ ν_A A and ν_B B yield just ν_c C and ν_D D.

For the total differential of the internal energy U and the enthalpy H follows

$$dU = (\delta U/dT)_{V,\zeta}dT + (\delta U/\delta V)_{V,\zeta}dV + (\delta U/\delta\zeta)_{T,V}d\zeta \tag{A32}$$

64 G. Wedler und H.J. Freund, "Lehrbuch der Physikalischen Chemie" Wiley-VCH, Weinheim 2012.

$$dH = (\delta H/\ dT)_{p,\zeta}dT + (\delta U/dV)_{T,,\zeta}dV + (\delta H/\delta\zeta)_{T,p}d\zeta \qquad \text{(A33)}$$

The differential quotients $(\delta U/\delta\zeta)_{T,V}$ and $(\delta H/\delta\zeta)_{T,p}$ describe the amount of heat, which is exchanged with the environment by one formula conversion in an isothermal and isochore reaction or in an isothermal and isobar reaction. However, the terms reaction energy and reaction enthalpy are preferred to the term heat of reaction.

Entropy

So far we were dealing with the internal energy of a system and the exchange of heat and work under different conditions. The result of our reflection is that energy can neither be lost nor gained, considering the system and the environment together it reveals that the content of energy is constant.

The first law of thermodynamics states that the energy of the universe is constant. This law results from our experience.

Another remarkable observation is that spontaneous processes always have one direction and the reverse direction is obviously not possible. Some aspects of the world determine the direction of spontaneous changes. A hot body and a cold body in contact equilibrate the temperature of both bodies, but it has never been observed that the uniform temperature of these bodies splits into a cold and a hot body. Another example is the expansion of a gas. An ideal gas expands spontaneously from a volume V_1 into an empty volume V_2 yielding the final volume $V_1 + V_2$ but it will never voluntarily compress $V_1 + V_2$ to V_1; this happens only when an outside agent does work.

What determines the direction of spontaneous change?

It is not the total energy, because the total energy of the universe – the system plus the environment in the thermodynamic sense – is constant.

Is it the energy of the system itself that tends toward a minimum? An ideal gas expands spontaneously into a vacuum, yet its internal energy does not change. Moreover, if the energy would decrease during a spontaneous process, the energy of the surrounding must increase by the same amount in order to satisfy the first law of thermodynamics.

When a change occurs, the total energy remains constant, but it is parceled out in different ways. Can the spontaneous change be related to some aspects of the distribution of energy? Let us look to the direction of change that leads to chaotic dispersal of the total energy.

As an example we concentrate on an isolated monoatomic gas. It has the simplifying feature that we have to consider only the chaotic, kinetic energy of translation of the atoms. Wherever an atom exists we will find kinetic energy. In the initial state, the gas in volume V_i after opening the connection to the larger volume expands to the final volume V_f. It is obvious that the probability to find a gas molecule in V_f is greater than to find it in V_i. We define a function called entropy, which is correlated with the probability of a state. The entropy of a system in the final state is greater than its entropy in the initial state. During a spontaneous change of state the entropy increases. This statistical interpretation of the entropy is clear.

How is entropy defined in the frame of classical thermodynamics?

We first have to distinct irreversible and reversible processes. The first are the spontaneous processes, they cause degradation in the quality of energy and hence increase of entropy in the surrounding. Reversible processes are finely balanced where the system is in equilibrium with its environment at every time. Every infinitesimal step along a reversible path is reversible. Reversible processes do not generate entropy but may transfer it from one part of the environment to another.

Based on observations, it is a general experience that all the spontaneous in the nature occurring processes always develop in the same direction. The system changes from a defined initial state into a defined final state. This final state is the equilibrium state of the system. In this state, the spontaneous process comes to a stop. In the equilibrium state, the process can be guided in both directions by infinitesimally small changes of the state parameters (p,T,V). The process can be reversed infinitesimally small steps without changes in the environment (reversible change of state).

I agree with the reader that the term "reversible change of a state" sounds artificially and indeed this is not observed in real nature, because it would take infinite time to come from the initial state to the final state. However, it was valuable to find the thermodynamic formulation of entropy.

From the "Carnot process"[65] one derives a function Q/T, the reduced heat. This leads to the introduction of a new function of state, called entropy S. It has been defined by Clausius to be

$$dQ_{\mathrm{rev}}/T = dS \tag{A34}$$

The law of entropy is the second law of thermodynamics; it is formulated as follows:

If the value of entropy of an isolated system at a given time has a distinct value, it remains constant if at a later time only reversible changes of state in the system occur. However, if in the change of state only one irreversible step is involved, the entropy of the system increases.

Water boils at 100° C at normal pressure. To evaporate the liquid, we heat it at a constant temperature. This heat of evaporation is entropy. The heat of melting is another example. Ice melts at 0° C, and the crystalline state is transferred to a liquid state, a system of lower order.

Like other functions of state, the entropy depends on variables of state. This is given by

65 See textbooks of physical chemistry, for example, G. Wedler and H.-J. Freund, "Lehrbuch der Physikalischen Chemie", Wiley-VCH, Weiheim 2012, p. 61ff.

$$S = f(p, T, n_i) \quad \text{and} \quad dS = (dS/dp)_{T,n_i +} + (dS/dT)_{p,n_i} + (dS/dn_i)_{p,T,n \neq ni} \tag{A35}$$

$$S = f(V, T, n_i) \text{ and } dS = (dS/dV)_{T,n_i +} + (dS/dT)_{V,n_i} + (dS/dn_i)_{V,T,n \neq n_i} \tag{A36}$$

A thermodynamically stable equilibrium state is reached by decreasing the energy and by increasing the entropy. This is the reason why two new functions of state have been defined. They have been introduced by Helmholtz and Gibbs: the Free energy A and the Free enthalpy G:

$$A = U - T \cdot S \tag{A37}$$

$$G = H - T \cdot S \tag{A38}$$

With these functions we get new conditions for the existence of an equilibrium in a closed system

$$dA_{V,T} = 0 \text{ and } dG_{p,T} = 0 \tag{A39}$$

if a spontaneous process occurs the following relations are valid:

$$dA_{V,T} < 0 \text{ and } dG_{p,T} < 0 \tag{A40}$$

Processes where the Free energy A and the Free enthalpy G increase are thermodynamically not possible.

In the following we focus on the Free enthalpy G, because in electrochemistry generally isobar and isothermic processes are of interest:

$$dG = (\delta G/dT)_{p,\zeta}\, dT + (\delta G/\delta V)_{T,\zeta}\, dV + (\delta G/\delta \zeta)_{T,p}\, d\zeta$$

Since p and T are constants, $dG_{p,T}$ is only a function of ζ, that is, the Free enthalpy of reaction.

In a simple redox reaction dG, respectively ΔG is given by

$$\Delta G = -n \cdot F \Delta E \tag{A41}$$

the relation that has already been given earlier in Chapter 2.

Appendix B

How to describe the electrical field

The relation between field strength and charge density

This chapter deals with the interrelation between electrical field strength, electrical potential and the space charge density. It will give a good basis for deeper insight in elementary concepts of electrochemistry.

The basic units for current, charge and tension (ampere, coulomb, volt) and the units for resistance and capacity (ohm, farad), deduced from the basic units, are arbitrary units, which have been chosen so that they can be connected to historical units by simple conversion. The historical units are based on the law of Coulomb, which in analogy to Newton's law in mechanics gives evidence on the force acting between two point charges.

This basic equation for the electrostatic interaction of two point charges q_1 and q_2 separated by a distance r in vacuum is

$$\mathcal{F} = 4\pi\varepsilon_0 \cdot q_1 \cdot q_2/r^2 \tag{B1}$$

where ε_0 is the vacuum permittivity ($8.854 \cdot 10^{-12}$/J C^2/m); the force F is a vector and has the dimension Newton; the charge is given by Coulomb C and the distance r by meter m.

One observes an interaction between two electric particles: a positive charged particle attracts a negative charged particle and vice versa. Particles with the same sign of charge repel each other. The direct conclusion from this observation is: A force exists, which is responsible for the interaction of electrically charged particles. This electric force is called electric field strength $\mathcal{E}$. Coulomb's law can be written as

$$\mathcal{F} = \mathcal{E}\,\mathrm{e} \tag{B2}$$

Obviously, electrical charges are surrounded by an electrical force field responsible for the interaction of charged particles.

How can we describe this scenario?

If we assign to every point in the space a vector of the electrical field strength we get a field of vectors. The interaction of the force field with the unit charge e is shown by the length and the direction of the vector. This represents the electric interaction at a point of the force field. By setting vectors at every point in space the electric force field around an electric charge can be described.

Characterizing a complex electrical field point by point using vectors may be somewhat confusing. Therefore, the vector field is imaged by lines of force. The lines of force are special curves in that in every point the tangent on the curve points to the direction of the electric field strength E.

As an example, the radial symmetric electric field around a charged sphere and the homogeneous field of a charged metal plate are shown in Fig. B1.
The strength of the field given by the length of the E vector can be presented by the density of lines of force, which is defined by the quotient:

$$\Phi = \text{Number of lines of force/perpendicular crossed area } A \tag{B3}$$

where Φ is called the flux of force. Φ depends on the field strength E and on the size and orientation of the area A. The size of the surface is shrinked to a differentially small size, and the orientation is expressed by a vector df, oriented perpendicular to the surface F under consideration.

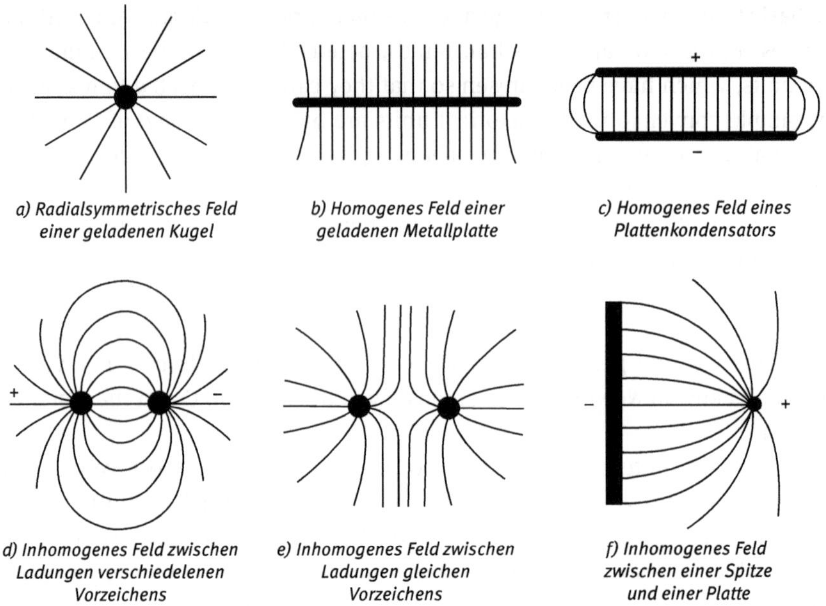

Fig. B1: Lines of force for different electrical fields.

Thus the flux of force through this element of the area A is given by the scalar product:

$$d\Phi = (\mathcal{E}\, df) = E\, df\, \cos\, \alpha \tag{B4}$$

where α is the angle between the vectors $\mathcal{E}$ and f.

The origin of a line of force is a positive electric charge, and the sink of a line of force is a negative electric charge. Lines of force connect positive and negative charges.

The force flux ϕ crossing an arbitrary surface is then given by the surface integral

$$\phi = \int^{F} (\mathcal{E} \cdot df) \tag{B5}$$

We define now in an electrical field a volume V which is limited by a closed surface F with arbitrary shape. If there is no electric charge inside the volume V, the total force flux is zero:

$$\phi = \int^{F} (\mathcal{E} \cdot df) = 0 \tag{B6}$$

because there is an equal number of ingoing and outgoing lines of force.

However, if the volume V contains the electric charge q, the following equation is obtained:

$$\phi = \int^{F} (\mathcal{E} \cdot df)\, k \cdot q \tag{B7}$$

Now we have to determine the proportional factor k.

We look at the specific case of the radial symmetric field of a single point charge in a sphere of the radius r. The density of the lines of force is at the surface of the sphere everywhere the same for a given r. E is constant and $\cos \alpha = 1$. Thus, the following relation is valid:

$$\phi = |\mathcal{E}| \cdot 4\pi r^2 = k \cdot e \tag{B8}$$

and with

$$\mathcal{E} = e/r^2 \tag{B9}$$

we get

$$k = 4\pi \tag{B10}$$

This allows the formulation of a general law:

$$\phi = \int^{F} (\mathcal{E} \cdot df)\, 4\pi\, e \tag{B11}$$

The total flux ϕ of force across a closed surface corresponds to the 4π-fold of the charge q in the volume surrounded by this surface

$$\phi = 4\pi\, q \tag{B12}$$

There are many systems where the charge is distributed over an area or in a volume, for example, electrolytes. We get space charge clouds with the charge density ρ (e.g., C/cm^3). The electrical charges in a volume V are given by eq. (B11) and can be written as

$$\phi = 4\pi \int^{V} S\, dV \tag{B13}$$

Since the electrical charges are the sources of the vector field, the surface integral over the closed surface is a measure for the “productivity” of the enclosed volume. Dividing by V leads to the “average productivity” of the sources in the volume V. For a differential small volume, that is, a point in space, the “productivity” of the source in vector analysis is called divE, that is, the divergence of the field strength. After further rearrangement finally results the equation

$$\mathrm{div}\mathcal{E} = \partial\mathcal{E}_x/\partial x + \partial\mathcal{E}_y/\partial y + \partial\mathcal{E}_z/\partial z = 4\pi s \tag{B14}$$

Equation (B14) is important. It formulates the relationship between field strength and charge density; if E is given as the function of the space coordinates, the size and the spatial distribution of charges are determined and vice versa giving the charge density defines the field strength.

The electrical potential

The description of the electrical field given above is sophisticated; therefore, electrochemists prefer a second, simpler method to describe the electrical field. They use the electrical work W resulting from the shift of a probing charge from one point P_0 to a point P_1 in the electrical field.[66] W is defined by the scalar product of the force $\mathcal{F}$ and the length of the distance s between P_0 and P_1:

$$dW = -(\mathcal{F}\,ds) = -F\cos\alpha\,ds \tag{B15}$$

where α is the angle between the acting force $\mathcal{F}$ and the direction of the shift s. To be in accordance with the definition in thermodynamics, energy, which is taken from outside, leading to a shift in the charge against the electric field is positive, and work taken from the electric field is negative.

The force $\mathcal{F}$ is connected with the field strength $\mathcal{E}$ by

$$\mathcal{F} = e/r^2 = \mathrm{E} \tag{B16}$$

This leads to the relation

$$dW = -e(\mathcal{E}\,ds) = -e\,|\mathcal{E}|\cos\,\alpha\,ds \tag{B17}$$

where $|\mathcal{E}|\cos\alpha$ is the component $\mathcal{E}_s$ of the field strength in the direction of s.

Shifting an electric charge over a finite distance from P_0 to P_1 yields the electrical work W:

$$W = -e\int_{P_0}^{P_1}(\mathcal{E}\,ds) \tag{B18}$$

W is independent from the way how e is shifted from P_0 to P_1. It is a “reversible work.”

If point P_0 is a fixed point and the integral is extended to variable limits with the coordinates x, y, z, the shift of $e = 1$ from P_0 to P_1 is

66 Gustav Kortüm “Lehrbuch der Elektrochemie”, Verlag Chemie, Weinheim 1966, p. 22.

$$-\int_{P_0}^{P_1} (\mathcal{E}\, ds) = \varphi_{x,y,z} - \varphi_{x_0,y_0,z_0} = \varphi_{x,y,z} - \text{const} \tag{B19}$$

If P_0 is located in infinity, φ becomes a scalar function called potential. Using the potential φ, the electrical field may be characterized point by point by the energy which results when we bring the probing charge $q_1 = 1$ from infinity via a distance s to the point $P_{x,y,z}$. Thus, the electrical field is characterized by a scalar field. The difference of two potentials φ_1 and φ_2 corresponds to the electric voltage U between these points.

It has to be mentioned that only the product of charge and potential $e \cdot \varphi$ has the dimension of energy. Due to the fact that e =1 sometimes the potential is virtually taken to be an energy.

So far we were dealing with point charges, now we have to go one step further on.

If we go from a defined point where the potential has a value φ via a distance ds to another point in the electric field, then the amount $d\varphi/ds$ depends on the direction, which we chose. There are many different possible values of the change $d\varphi/ds$, but in one direction the change in the potential is a maximum. This maximum change constitutes a vector in vector analysis called the gradient of φ. Changes in other directions can be related to the gradient by the scalar product of the gradient and the arbitrary direction ds:

$$(\text{grad}\, \varphi\, ds) = \text{grad}\, \varphi\, \cos \varphi \cdot \cdot ds \tag{B20}$$

α is the angle between both the vectors. For parallel vectors ($\cos \varphi = 1$) dφ reaches its maximal value. Integration over a finite distance between P_1 and P_2 leads to the relation

$$\int_{P_1}^{P_2} (\text{grad}\, \varphi \cdot ds) = \varphi_2 - \varphi_1 \tag{B21}$$

Comparing this relation with eq. (B19) leads directly to

$$\mathcal{E} = -\,\text{grad}\, \varphi$$

The electric field strengths $\mathcal{E}$ can be presented by the gradient of the scalar function φ of the electrostatic potential.

Connecting points of the same values of the potential we get equipotential surfaces, on which the vector $\mathcal{E}$, respectively grad φ, is oriented perpendicularly. Therefore, from the equipotential surfaces the direction of the electric field strength can be directly evaluated. As an example, Fig. B1 shows the equipotential surface of a point charge; the electric field is radially symmetric, and the equipotential surfaces are spheres. The field strength is inversely proportional to the distance of neighboring equipotential surfaces. This corresponds to the presentation of the electrical field by lines of force to the density of lines of forces.

The interaction between point charges is of basic interest, however, in electrochemistry we consider huge numbers of charges in a limited space or area, that is, the space charge density. The space charge density is given by the number of electric charges per unit volume.

A very important relation is the differential equation of the electrostatic potential. It follows by combining equations div **E** = $4\pi\rho$ and **E** = – grad ψ:

$$\text{div grad}\,\psi = -4\pi\rho \tag{B22}$$

This is the Poisson equation. It connects the electrostatic potential with the charge density .In cartesian coordinates the potential equation reads:

$$\text{grad}\,\psi = \psi\;(\delta\psi/\delta x)\,\mathbf{i} + \psi\;(\delta\psi/\delta y)\,\mathbf{j} + \psi\;(\delta\psi/\delta z)\,\mathbf{k} \tag{B23}$$

i, j, k are the unit vectors in the *x*-, *y*-, *z*-directions:

$$\text{div grad}\,\psi = \delta^2\psi/\delta x^2 + \delta^2\psi/\delta y^2 + \delta^2\psi/\delta z^2 = \Delta\psi = -4\pi\rho \tag{B24}$$

Δ in this equation is the Laplace operator.

The Poisson equation is used in the Debye–Hückel model of electrolytes and in the description of the diffuse double layer. For application of the Poisson equation, see Box 4.

Appendix C

Galvani potential, Volta potential and surface potential

In electrochemistry, the terms potential and electrochemical potential are central concepts. It seems appropriate to evaluate these terms in some detail. In the following, we follow mainly the considerations of E. Lange[67] described by H.-J. Freund.[68]

We look at a metallic sphere and an electrical test charge. In metals the metal ions are arranged in a crystal lattice; frequently this is cubic close packed. Between the metal ions exists the "electron gas." It is generated by removal of the outermost valence electrons from the metal atoms. The electrons in the electron gas are very mobile. The huge number of their electronic states leads to the formation of energy bands, and the band structure is characteristic for each metal. The potential in the interior of the sphere referring to the origin is the "inner potential" or "Galvani potential" φ. In the interior of the metal phase a test charge is on the Galvani potential φ.

At the edge of the metal, the phase boundary, the forces that influence the ions and the electrons are not symmetric. At the boundary of metals the electron gas

67 E. Lange, Professor of physical chemistry at the University Erlangen.

68 Gerd Wedler und H.-J. Freund Lehrbuch der Physikalischen Chemie, Wiley-VCH Verlag 2012.

exceeds the surface which is defined by the metal ions. Based on this, at the boundary of the metal a dipole layer is generated. This is the reason for an additional potential at the surface and the surface potential χ.

A test charge ($q = 1$) in infinite distance from the metal is not influenced by electrical forces (Coulomb law: $K = Q \cdot q / 4\pi\varepsilon_0 r^2$); therefore, the origin ($\varphi_\infty = 0$) of the potential scale is defined by the electrical potential φ_∞ in infinite distance from the metal. The energy of the test charge q at infinity is zero. The electrical potential changes in the direction of the metal. Shifting of the test charge in direction to the metal changes their potential energy $q \cdot \varphi$. At about 10^{-6} cm in front of the surface the potential energy reaches its maximum value, but it is not influenced by the dipole layer. The potential at this point is called "outer electrical potential" or Volta potential ψ. Transferring the test charge into the interior of the phase to the Galvani potential φ one has to overcome the surface potential χ. Therefore, the inner potential φ can be formulated by the sum of the Volta potential ψ and the surface potential χ:

$$\varphi = \psi + \chi \tag{C1}$$

This is a general relation, but the surface potential may be modified by adsorption of molecules: χ is not a quantity specific for a material. Even at surfaces free of adsorbates χ depends on the arrangement of atoms and depends therefore on the indication of the surface symmetry of a crystal.

The Volta potential can be measured, in contrast to the Galvani potential and the surface potential which cannot be measured. The reason is that electrical charge in the form of an electron or ions is always bound to matter. The transport of charge across the phase boundary involves changes in the surrounding and therefore a chemical work appears besides the electrical work. These different energies are not distinguishable, and can only be measured together.

Transferring 1 mol of electrons or ions with the charge z_i, as described earlier, into the interior of a phase we have to take into account the electrical and the chemical work.

Therefore, in analogy to the chemical potential μ_i, the "electrochemical potential" $\mu^e{}_i$ and the "real potential" α_i is defined for ions and the electron and generally for electrically charged particles:

$$\mu^e{}_i = \mu_i + z_i F \cdot \varphi \tag{C2}$$

$$\alpha_i = \mu_i + z_i F \cdot \chi \tag{C3}$$

$$\alpha_i = \mu_i - z_i F \cdot \psi \tag{C4}$$

For following considerations, it is necessary to look at the terms Galvani voltage $\Delta\varphi$ and Volta voltage $\Delta\psi$. The term scheme of metal electrons (see Appendix D) takes into account only the electrons of the conduction band, and inner shell electrons are not regarded. The origin of the energy scale ($\mathcal{E} = 0$) is again the energy of the motionless electron in the interaction-free vacuum. In the metal, the energy ε_0 of the electrons consists of three contributions. The electrical part $- e \cdot \varphi_0$: If the metal has no excess charge, it is given by the surface potential, that is, by $- e \cdot \chi$; the binding energy $- E_{pot}$ on the positive atom ions and of the Fermi energy E_F, which represents the degree of filling of the conduction band with electrons:

$$E_0 = - e\chi - E_{pot} + E_F \tag{C5}$$

On the other hand, this is the energy that has to be added to the electrons if they are ejected into the vacuum. E_0 is identical with the work function. Related to one electron this is the real potential α_0/N_A or the electrochemical potential $\mu^e{}_0/N_A$. Then the chemical contribution $(-E_{pot} + E_F)$ and the chemical potential μ/N_A are related to one electron. All the energies E_0, $e\chi$, E_{pot}, and E_F have different values for different metals. This is taken into account for the metal A and metal B in an energy scheme.

When the metals A and B are in touch and $E_A \neq E_B$, electrons flow from the metal where they have the higher energy to the metal where they have the lower energy. In our example, they flow from metal A to metal B. This transfer lasts until the energies of the electrons are the same in both metals, $E_{K,A} = E_{K,B.}$ By the electron transfer the metal B gets a negative excess charge, whereas the metal A gets a positive excess charge. These result in two oppositely directed outer potentials ψ_A and ψ_B. At the phase boundary exists, developed by electron transfer, a Galvani potential $\Delta\varphi_K$ and a Volta potential, respectively, contact potential $\Delta\psi_K$. Tantamount to this is the statement that the Fermi edges in relation to the vacuum level become equivalent.

Linking eqs. (C3) and (C5) and taking into account $z_i = -1$ leads to

$$\alpha_{K,A} - F\psi_{K,A} = \alpha_{K,B} - F\psi_{K,B} \tag{C6}$$

and

$$F(\psi_{K,A} - \psi_{K,B}) = \alpha_{K,A} - \alpha_{K,B} = \alpha_{0A} - \alpha_{0,B} \tag{C7}$$

because α_A and α_B are not changed by the contact of the two metals (μ and χ remain constant). The product of $\Delta\psi_K$ and the Faraday constant F is the difference of the real potential of both the metals

$$F \cdot \Delta\psi_K = \Delta\alpha_0 \tag{C8}$$

that is, the product is equivalent to the negative difference of their mass referring work function.

$e\varphi_K$ is given by

$$e\varphi_K = e\chi + e\psi_K \tag{C9}$$

and therefore

$$F(\varphi_{K,\mathrm{A}} - \varphi_{K,\mathrm{B}}) = F(\chi_{K,\mathrm{A}} - \chi_{K,\mathrm{B}}) + F(\psi_{K,\mathrm{A}} - \psi_{K,\mathrm{B}}) \tag{C10}$$

and considering eq. (C7) follows

$$F\Delta\varphi_K = F\Delta\chi + F\Delta\alpha_0 \tag{C11}$$

The Galvani potential $\Delta\varphi_K$ which results from the contact of two metals may be calculated from the difference in the surface potentials and the work functions of the metals referring to their amount of matter.

Appendix D

The electronic structure of semiconductors

Semiconductor electrochemistry has forced the collaboration of solid-state physicists with electrochemists.[69] From this collaboration evolved a common terminology for describing the semiconductor/electrolyte interface. The convention is largely based on the energy band model of solid-state physics. All events are described in terms of charge moving between energy levels. A shortcoming is that there is no way describing chemistry at the surface.

The forthcoming text will be qualitative, only a few equations will be given without any derivation.

The electronic structure of solids

Solids and especially semiconductors of interest are crystalline networks build up from atoms.[70] The covalent bond network of these atoms extends throughout the volume of the crystal. As required by molecular orbital theory, orbitals of neighboring atoms split into a binding term and an antibonding term. In a macroscopic solid, splitting occurs in a very large number of closely spaced terms. This ensemble of energy levels is called an energy band. The energy spacing between adjacent energy levels is so small that energy bands are treated as a continuum of energy levels. This continuum has well-defined upper and lower energetic limits.

---

69 This chapter follows essentially “Semiconductor Electrodes“ ed. Harry O. Finklea, Elsevier Amsterdam 1988.

70 Roald Hoffmann, "Solids and Surfaces, A Chemists View of Bonding in Extended Structures" VCH Verlagsgesellschaft, Weinheim 1988 (A highly recommended small booklet).

The energy-level diagram of a solid consists of energy bands, and some of them may overlap. The highest filled energy band and the lowest empty energy band are the most important for charge transport and charge transfer. All subsequent discussion will concern them.

In a semiconductor or an isolator, the bands are separated by a gap devoid of energy levels. In a metal the two bands overlap, or a band is partially filled with electrons. The highest filled band is called valence band and the lowest empty band is the conduction band. In most common semiconductors, the valence band is derived from a bonding molecular orbital and the conducting band from an antibonding molecular orbital. This has important consequences when charge carriers reach the semiconductor/electrolyte interface.

Figure D1 shows the essential features of the energy band diagram. E_c marks the lower limit of the conduction band and E_v the upper limit of the valence band, The band gap energy E_{bg} is simply the difference between E_c and E_v. The preferred energy unit for E_{bg} is electron volt (eV). This unit translates readily into the electrochemical energy scale.

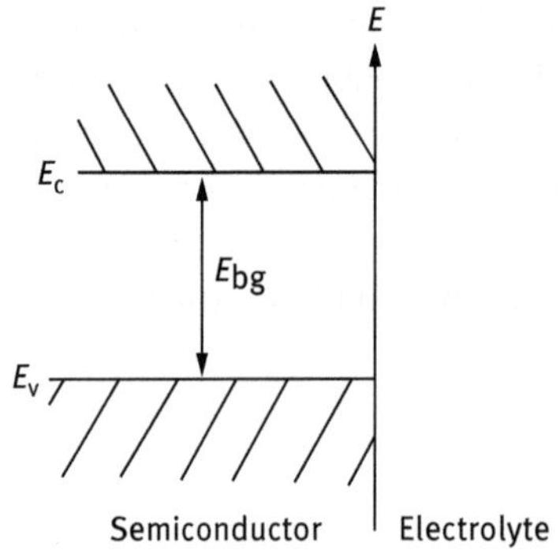

Fig. D1: Energy band diagram of an intrinsic semiconductor in contact with an electrolyte. E_c is the lower edge of the conduction band, E_v is the upper edge of the valence band and E_{bg} is the band gap energy.

The generation of charge carriers

A metal conducts electricity because it has empty energy levels so that electrons migrate immediately in the empty ones. Electrons are thermally excited to empty levels and acquire lateral mobility along the adjacent empty levels. A perfect semiconductor at absolute zero temperature is a nonconductor, because the valence band is completely occupied and the conduction band is totally empty. To impart electrical conduction, charge carriers must be created by one of three mechanisms: thermal generation, photoexcitation or doping.

Photons can excite electrons from the valence band to the conduction band. Two charge carriers are generated: the electron in the conduction band and the hole, a positive vacancy in the valence band (Fig. D2). Both are mobile within their respective energy bands.

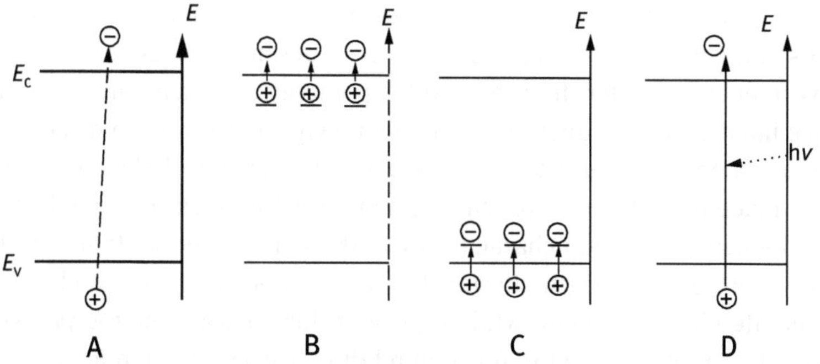

Fig. D2: Mechanisms of charge carrier generation in a semiconductor. (A) Thermal generation; (B) *n*-type doping; (C) *p*-type doping; and (D) photoexcitation.

The concentration [*n*] of electrons and holes [*p*] depends on the absolute temperature and the band gap energy:

$$[n][p] = N_c N_v(-E_{bg}/kT) \tag{D1}$$

N_c and N_v are the "effective" density of energy levels at the conduction band edge and the valence band edge, respectively. At room temperature their value is of the order of $10^{19}/\text{cm}^3$. When thermal generation is the only mechanism of charge carrier generation, as in an intrinsic semiconductor, then [*n*] = [*p*], and the total charge carrier concentration (conductivity) drops rapidly with increasing band gap energy for a given temperature. At room temperature (25 °C), *kT* is 0.0257 eV. Consequently, undoped (intrinsic) semiconductors with a band gap energy greater than 0.5 eV are essentially insulators.

Doping provides a meaning of enhancing conductivities of wider band gap semiconductors. In doping, energy levels are introduced into the band gap region by chemically altering the crystal lattice. The crystal can be made nonstoichiometric so that one element is in excess or is deficient, or elements of different valencies can be substituted into the crystal lattice. The new energy levels must be near one of the band edges. Two cases arise. When the doping entity is an electron donor and is located a few kT of the conduction band edge, then a large population of electrons is donated into the conduction band by thermal excitation. Equation (D1) still applies; consequently, the hole concentration drops when the electron concentration increases. Electrical current is carried predominantly by the negative charge carriers in the conduction band. This semiconductor is called *n*-type.

The corresponding *p*-type semiconductor is created when electron acceptors have energy of e few kT of the valence band edge. Electrons are captured from the valence band creating an excess population of positive charge carriers, that is, holes. By means of doping useful conductivity can be imparted in semiconductors with energy gaps up to 3.5 eV.

The presence of nearly filled energy bands separated from nearly unfilled energy bands leads to the very interesting photosensitivity of semiconductors.

A photon with energy greater than the band gap energy can excite an electron from the valence band to the conduction band. The newly created hole and the excited electron may recombine, but they can be separated by means of electric fields or diffusion, and made to perform useful chemical or electrical work. The band gap energy controls the light absorption characteristic of the semiconductor. In terms of wavelength, light with a wavelength shorter than a threshold wavelength (λ_{bg}) is absorbed to generate charge carriers, while light with longer wavelengths passes unimpeded through the semiconductor. A useful relation between λ_{bg} and E_{bg} is

$$\lambda_{bg} = 1,240/E_{bg} \tag{D2}$$

where λ_{bg} has units of nanometers and E_{bg} has units of electron-volts. The solar energy conversion field of sun light into electricity (or fuel) has been estimated to be 1.4–1.8 eV.

The Fermi level

The Fermi level is an extremely important parameter for metal and semiconductor electrochemistry, because it is the property that is controlled by the externally applied potential. Thermodynamically, the Fermi level is the electrochemical potential of the electron in the solid. An equivalent definition arises from the distribution of electrons among energy levels in a solid: the Fermi level is the energy (E_f) at which the probability of an energy level being occupied by an electron is exactly ½. This probability of occupancy increases with energies below the Fermi level and decreases with energies above the Fermi level in accordance with the Fermi–Dirac distribution function:

$$f(E) = (1 + \exp(E - E_f/kT))^{-1} \tag{D3}$$

In metals, the Fermi level in effect marks the division between occupied and empty energy levels. In semiconductors, the Fermi level resides in the bandgap region. The word an actual energy level is misleading in that it implies the existence of an actual energy level, but neither the thermodynamic nor the statistical definition requires one.

Where the Fermi level resides relative to the band edges depends on the doping state, for an intrinsic semiconductor the Fermi level is approximatively midway between E_c and E_v.. This equation is required by equation (D3) since the valence band is essentially completely filled ($f(E)$ —> 1 if $E_v \ll E_f$) and the conduction band is nearly empty ($f(E)$ —> 0, if E_c E_f). For *n*-type doping, the Fermi level shifts toward the conduction band edge while for *p*-type doping the Fermi level shifts toward the valence band edge. The position of the Fermi level is related to the doping level (charge carrier concentration).

For electrochemical work, semiconductors are often doped to charge carrier concentrations of 10^{18}–$10^{19}/cm^3$. Consequently, the Fermi level shifts to a position within 0.1 eV of a band edge. In such a situation the band edge and the Fermi level are often assumed to be at the same energy. If doping levels are even higher, the Fermi level moves into one of the energy bands and the semiconductor becomes degenerate, that is, a quasi-metal.

At equilibrium, the Fermi level is at constant energy throughout the solid; otherwise, charge will flow until E_f is isoenergetic. The rapid rate of charge generation and charge flow within a conducting solid implies that semiconductors connected to metals (or other semiconductors) will have a common Fermi level.

Application of the equilibrium concept to the electrode/electrolyte interface requires the definition of the Fermi level in an electrolyte. An electrolyte that contains both forms of a redox couple has a "Fermi level" identified with the Nernst potential E:

$$E = E_0 - (RT/nF)\ \ln\ a(\text{Red})/a(\text{Ox})E \tag{D4}$$

$$E_{\text{redox}} = e\ E \tag{D5}$$

E_{redox} is the "Fermi level" of the solution expressed on the energy scale, whose reference point is defined below. The solution "Fermi level" arises from the concept that "Ox" and "Red" molecules in an electrolyte form a set of electron energy levels, of which the "Ox" molecules are the unoccupied levels and the "Red" molecules are the occupied ones.

When an otherwise unconnected electrode, a single electrode, is immersed into an electrolyte containing a redox couple (both forms must be present). The "Fermi level" of the electrode shifts with charge flow across the interface until $E_f = E_{\text{redox}}$. The rate at which equilibration takes place depends critically on the rate of charge transfer across the interface. In many important cases, the rate of equilibration is much slower for a semiconductor than for a metal.

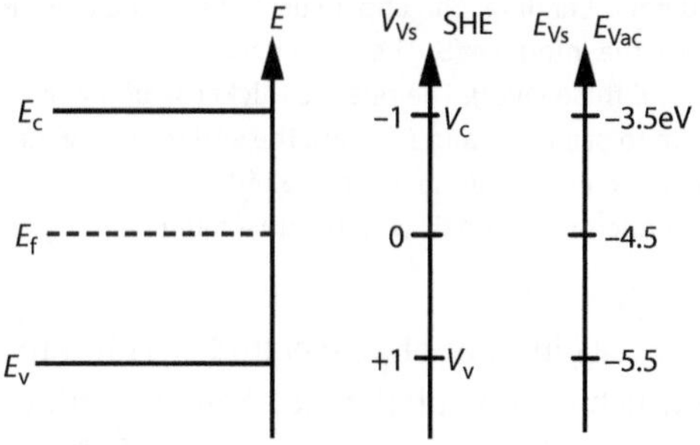

Fig. D3: The energy scale for a semiconductor and the potential scale for an electrode. Intrinsic semiconductor with $E_{bg} = 2$ eV.

The two energy scales (Fig. D3) are related by the theoretical value for the Fermi level of the standard hydrogen electrode (SHE):

$$V = e(-4.5\ V - E) \tag{D6}$$

By convention Fermi level energies are defined relative to the energy of the electron in vacuum. The potential scale is defined relative to the SHE, although in experiments mostly different reference electrodes are used. The potential scale increases whereas the energy scale of the semiconductor decreases; electrons spontaneously flow from the more negative potential to a more positive potential.

Knowing the electrode potential and the position of the Fermi level relative to the band edges, it is possible to locate the band edges of the semiconductor on the potential scale.

The effects of applied potentials

We have a look at an electrochemical cell consisting of a semiconductor electrode and a nonpolarizable reference electrode, with both are immersed in an electrolyte.

An externally applied potential controls the Fermi level of the semiconductor electrode with respect to the Fermi level of the reference electrode. We know from preceding chapters that there exist potential changes between the working electrode and the reference electrode, localized near the respective interfaces. All changes in potential affect the potential drop across the semiconductor/electrolyte interface. At a fixed potential there are three regions of varying electrical potential (Fig. D4).

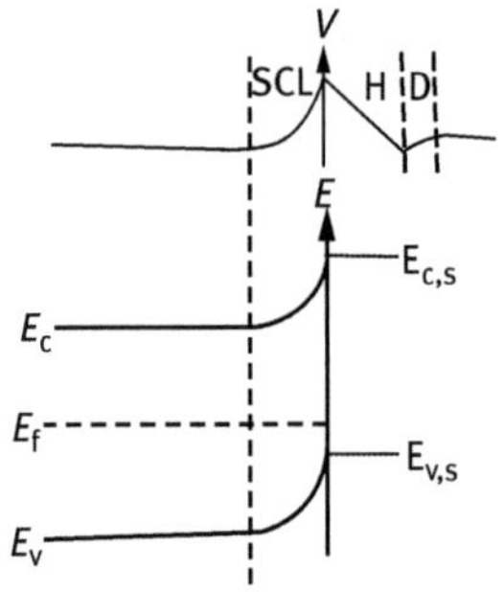

Fig. D4: Electrical potential drop across the semiconductor/electrolyte interface and band bending. The upper curve shows the electrical potential across the interface (SCL: space charge layer, H: Helmholtz layer, D: diffuse layer). The relative thickness of these layers is not shown to scale. E_{cs} and E_{vs} mark the surface energies of the conduction band edge and the valence band edge, respectively. The semiconductor is intrinsic with positive band bending.

In the electrolyte, a well-known structure exists in front of an electrode. Within the electrode there is a space charge layer in which the potential varies from the surface value to the bulk value. The term "space charge" refers to the presence of charge occupying the space near the surface; the charge results from the charged doping sites or the accumulation of holes or electrons at the surfaces. The concentration of holes and electrons is perturbed at the surface relative to the bulk. Remembering the Poisson equation, of course, one expects that an excess charge close to the surface leads to a change in the potential

The Fermi level stays at constant energy from the bulk to the surface, so the position of the Fermi level relative to the band edged changes at the surface.

A change in the externally applied potential results in a change in the potential drop across both sides of the interface, but in certain cases the change is confined to the space charge layer in the semiconductor. In accordance with the double layer theory, the side at the interface containing the lower concentration of charge carriers experiences the grater change in potential, when the applied potential is changed. The typical doping level of a semiconductor, 10^{18}–10^{19} charge carriers/cm^3, is considerably less than the concentration of ions in a typical electrolyte (10^{22} for a 1:1 salt of 0.1 M concentration). Consequently, the band bending of the semiconductor is directly affected by externally applied potential.

There exists for semiconductors as well as for electrolytes a unique potential for which the potential drop between the surface and the bulk of the electrode or electrolyte, respectively, is zero. This is the potential of zero charge (PZC) at a metal/electrolyte interface; in analogy there exists a potential where at the surface of the semiconductor no excess charge exists. This potential is called flat band potential V_{fb} since the band edged to not bend. When the electrode is biased positive of the flat band potential, the Fermi level is lowered and some fraction of $V–V_{fb}$ appears across the space charge layer. Likewise, potentials negative of V_{fb} cause downward band bending (Fig. D5).

If the semiconductor is intrinsic (Fig. D5) and relatively free of surface states (energy levels on the surface and in the bandgap region) then all of the potential change relative to V_{fb} appears in the space charge layer. In this situation, the magnitude of band bending (E_{cb} – E_{cs}) in eV equals the change in the applied potential (V – V_{fb}) with respect to V_{fb}:

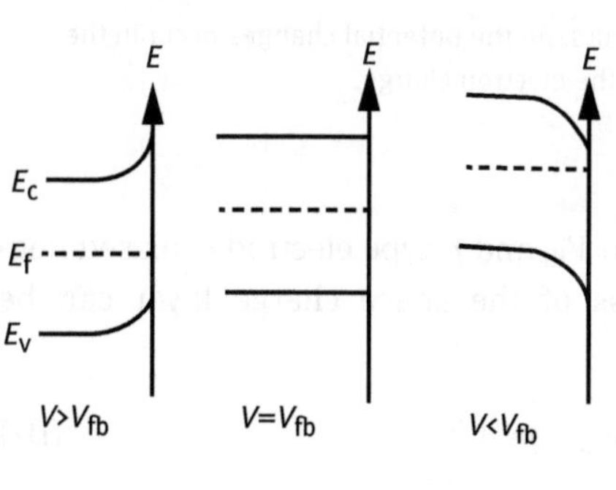

Fig. D5: Band bending as a function of the applied potential; V, applied potential; V_{fb}, flat band potential. Intrinsic semiconductor.

$$E_{cb} - E_{cs} = e(V - V_{fb})$$

Thus the direction of band bending can be predicted knowing the applied potential relative to V_{fb}.

This highly desirable condition is also obtained for n- and p-type semiconductors, provided that the applied potential moves the Fermi potential away from the nearest band edge at the surface (Fig. D6).

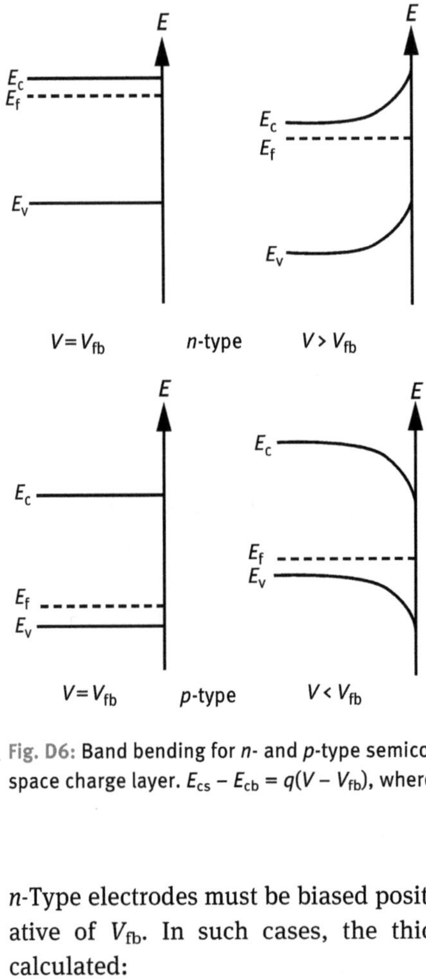

Fig. D6: Band bending for n- and p-type semiconductors. All the potential changes occur in the space charge layer. $E_{cs} - E_{cb} = q(V - V_{fb})$, where q is the electron charge.

n-Type electrodes must be biased positive to V_{fb} and p-type electrodes biased negative of V_{fb}. In such cases, the thickness of the space charge layer can be calculated:

$$W = [2\ \varepsilon\ \varepsilon_g (V - V_{fb})/qN_d]^{1/2} \tag{D7}$$

where N_d is the doping level, ε the dielectric constant in the direction normal to the surface and ε_d the permittivity of the free space (Fig. D7).

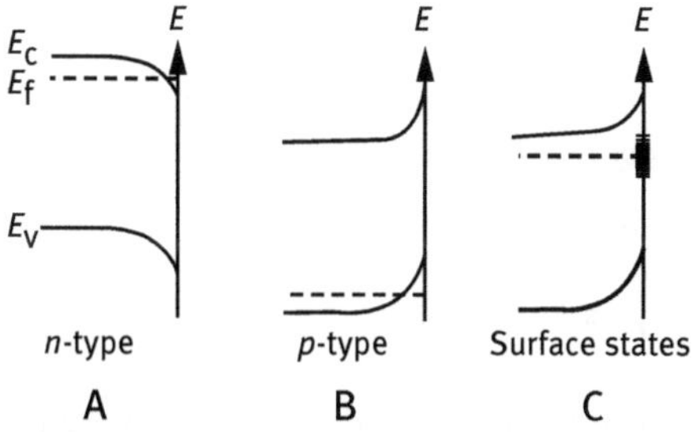

Fig. D7: Band bending leading to degeneracy in *n*-type (A) and *p*-type (B) semiconductors, and Fermi-level pinning by surface states. In all three cases, potential change occurs in both the space charge and the Helmholtz layer.

For further reading, the book *Semiconductor Electrodes* (ed. Harry O. Finklea, Elsevier 1988) is recommended.

Appendix E

Debye–Hückel theory

In this appendix, for the interested reader some more details are given for deriving the important results of this model.[71]

The leading equation is the Poisson equation formulated in polar coordinates

$$1/r^2 \cdot \delta/\delta r\,\{r^2 \delta\psi(r)/\delta r\} = -\rho(r)/\varepsilon\varepsilon_0 \tag{E1}$$

In the origin of the coordinates, the central ion *j* is located. The coordinate system is rigidly connected with the central ion and follows all movements of the central ion. In the surrounding volume element *dV* due to the electrostatic force counter-ions exist, developing a potential ψ and forming the ion atmosphere. The total charge of the ion atmosphere is opposite equal to the charge of the central ion.

Setting for the volume element under consideration $dV = 4\pi r^2 dr$ follows

$$\int_a^\infty 4\pi r^2 \rho dr = -\ z_j e_0 \tag{E2}$$

where *r* is the minimum distance between the central ion and other ions.

$N_1, N_2, \ldots, N_i$ is the number of different ions per cm^3 in the solution with the charges $e_1 = z_1 e_0$, $e_2 = z_2 e_0, \ldots, e_i = z_i e_0$. Taking into account the thermal energy and

71 G. Wedler, H.-J. Freund, "Lehrbuch der Physikalischen Chemie", Wiley-VCH 2012, p 229ff.

following the distribution law of Boltzmann the concentration of these ions in the volume element dV is

$$dN_i = N_i \exp\left[-z_i e_0\ \psi/kT\right] \tag{E3}$$

$$^1N_i(r)/^1\tilde{N}_i = e^{-zi\cdot e\cdot\psi(r)/kT} \tag{E4}$$

$$\rho(r) = \sum z_i\cdot e\cdot {}^1\tilde{N}_i\cdot e^{-zi\cdot e\cdot\psi(r)/kT} \tag{E5}$$

The next steps are some mathematical transformations. Provided we have very diluted solutions for which $z_i\cdot e\cdot\psi(r) \ll kT$ is valid, we can transform the e-function into the progression

$$e^{-zi\cdot e\cdot\psi(r)/kT} = 1 - [z_i\cdot e\cdot\psi(r)]/kT + 1/2!\ \{[z_i\cdot e\cdot\psi(r)]/kT\}^2 + \cdots \tag{E6}$$

and terminate the progression after the second member.

We get:

$$\rho(r) = \sum_i z_i\cdot e\cdot {}^1\tilde{N}_i[1 - z_i.e.\psi(r)]/kT] \tag{E7}$$

and

$$\rho(r) = \sum_i z_i\cdot e\cdot {}^1\tilde{N}_i - [e^2\cdot\psi(r)]/kT\ \cdot\sum_i z^2{}_i\cdot {}^1\tilde{N}_i \tag{E8}$$

An additional simplification of this expression is possible. The first term refers to the summation of all charges; because of the electroneutrality of the solution this term is zero.

We express the average number $^1\tilde{N}_i$ of ions by the concentration c_i and introduce the term “ionic strength“

$$^1\tilde{N}_i = N_A\cdot c_i \tag{E9}$$

and the ion strength I is defined by

$$I = 1/2\sum z_i{}^2\cdot c_i \tag{E10}$$

The density of charge can now be formulated as

$$\rho(r) = -\ [2\ N_A e^2\cdot I/kT]\cdot\psi(r) \tag{E11}$$

The Poisson equation is now

$$1/r^2\cdot\delta/\delta r\{r^2\,\delta\psi(r)/\delta r\} = [2N_A e^2\cdot I/\varepsilon\varepsilon_0 kT]\cdot\psi(r) \tag{E12}$$

We simplify the form of the equation by introducing the abbreviation $1/\beta$:

$$1/\beta = [2\ N_A e^2\cdot I/\varepsilon\varepsilon_0 kT]^{1/2} \tag{E13}$$

Finally, we get the equation

$$1/r^2 \cdot \delta/\delta r\,\{r^2\,\delta\psi(r)/\delta r\} = (1/\beta)^2 \cdot \psi(r) \tag{E14}$$

which is now called the Poisson–Boltzmann equation.

Solving this equation leads to

$$\psi(r) = A/r\; e^{-r/\beta} + B/r \cdot e^{1/\beta} \tag{E15}$$

where A and B are integration constants. They can be determined by using the electroneutrality and the condition: if $r \to \infty$ then $\psi \to 0$. This latter condition is fulfilled for $B = 0$. Thus the equation is reduced to

$$\psi(r) = A/r\; e^{-r/\beta} \tag{E16}$$

From the electroneutrality follows that the charge of the complete spherical ion cloud is equal to the charge of the central ion, but with opposite sign. The total charge of the ion cloud is received by integration of the charge density over the complete volume of the ion cloud. We assume the ion cloud to be spherical. The charge density dependent on the radius r is received by equation

$$\rho(r) = -\,\varepsilon \cdot \varepsilon_0 \cdot A/\beta^2 r \cdot e^{-r/\beta} \tag{E17}$$

via substitution of eq. (E16) into eq. (E11) and taking into account eq. (E13).

The total charge of the ion cloud is then given by$_{\text{i}}$

$$\int \rho dV = -\varepsilon \cdot \varepsilon_0 \cdot A \int_a^{\infty} 1/\beta^2 \cdot 1/r \cdot e^{-r/\beta} \cdot 4\pi r^2 dr \tag{E18}$$

The lower limit a of the integral is the radius of the central ion, and the ion cloud extends to infinity. Due to electroneutrality the left side of this equation corresponds to the charge of the central ion:

$$-z_i \cdot e = -\varepsilon \cdot \varepsilon_0 \cdot A \int_{a/\beta}^{\infty} 1/\beta \cdot 1/r \cdot e^{-r/\beta} \cdot 4\pi r^2 d(r/\beta) \tag{E19}$$

This is a defining equation for the constant A; by partial integration results $(1 + a/\beta)$ $e^{-a/\beta}$; finally we get A

$$A = z_i e/4\pi \cdot \varepsilon\varepsilon_0 \cdot e^{-a/\beta}/(1 + a/\beta) \tag{E20}$$

Substituting eq. (E20) into eq. (E16) we get the central equation of the Debye–Hückel theory:

$$\psi(r) = z_i e/4\pi\varepsilon\varepsilon_0 \cdot e^{-a/\beta}/(1 + a/\beta) \cdot e^{-r/\beta}/r \tag{E21}$$

Appendix F

Potential variation in the diffuse double layer

To describe the diffuse part of the double layer, the centers of attention are again the Boltzmann equation and the Poisson equation. A procedure is used comparable to that reported for the Debye–Hückel theory.

The local concentration of anions c^- and cations c^+ at a distance r from the electrode surface and the potential ψr is determined by the Boltzmann distribution

$$c^{\pm} = c_0^{\pm} \exp(-z^{\pm} e \psi(r)/kT) \quad \text{(F1)}$$

$c_0^{\pm}$ is the concentration of cations and anions in the bulk of the solution. It is connected with the corresponding charge $z^{\pm}$ by $c^+z^+=|c^-z^-|$, the principle of electroneutrality. Considering only the electrostatic interaction the charge density at the distance r is given by

$$\rho_r = z^+ e\, c_0^+ \exp(-z^+ e\, \psi(r)/kT) + z^- e\, c_0^- \exp(-z^- e\, \psi(r)/kT) \quad \text{(F2)}$$

In an isotropic dielectric medium, the Poisson equation connects the potential ψ at every arbitrary point with the charge density:

$$\varepsilon_r \nabla \phi = -4\pi \rho_r \quad \text{(F3)}$$

For the one-dimensional distribution in front of the infinite planar electrode the Poisson equation is given by

$$\delta/\delta r\, (\varepsilon_r\, \delta\psi_r/\delta r) = -4\pi \rho_r \quad \text{(F4)}$$

By substitution of ρ_r in eq. (F4) by eq. (F2) results the Poisson–Boltzmann equation:

$$\delta^2 \psi r/\delta r^2 = -4\pi e/\varepsilon \Sigma c_0^{\pm} z^{\pm} \exp(-z^{\pm} e\, \psi r/kT) \quad \text{(F5)}$$

In eq.(F4) one assumes that the permittivity ε_r can be substituted by the permittivity ε of the solvent and eq. (F5) can be integrated.

With the identity

$$\delta/\delta\psi_r\, (\delta\psi_r/\delta r)^2 = 2\delta^2 \psi_r/\delta r^2 \quad \text{(F6)}$$

and the boundary conditions

$$\delta\psi r/\delta r \to 0 \text{ if } r \to \infty \text{ and } \psi_r \to \psi_s \quad \text{(F7)}$$

(ψ_s is the potential in the interior of the solution when $\rho = 0$ due to electroneutrality) follows:

$$\delta\psi r/\delta r = \pm[8\pi kT/\varepsilon\Sigma c^{\pm}{}_0\{\exp(-z\pm e_0(\psi_r-\psi s)/kT)-1\}]^{1/2} \tag{F8}$$

From eq. (F8) one gets a relation between the field strength $\mathcal{E}$ and the potential in a distance r from the electrode. For symmetrical electrolytes:

$$|z^{\pm}| = z^{+} = |z^{-}| = z \text{ and } c^{\pm} = c^{+} = c^{-} = c \tag{F9}$$

A plausible reflection shows that only the negative root in eq. (F8) makes sense if specific adsorption can be excluded. After rearrangement of eq. (F8) follows

$$\delta\psi r/\delta r = -(32\pi\ kT\ c_0/\varepsilon)1/2\cdot\sinh[(z\ e_0(\psi_r-\psi_s)/2kT)] \tag{F10}$$

Using the law of Gauss

$$q^M = \varepsilon/4\pi\cdot\delta\psi/\delta r \tag{F11}$$

one can realize a relation between the area charge density q^M in eq. (F11) at the phase boundary metal electrolyte and the potential ψ_a at the minimal distance between the metal surface and the ions (eq. (F12))

$$q^M = -q^d = (2\ kT\ \varepsilon c_0/\pi)1/2\cdot sinh\ [(\ z\ e_0(\psi^r-\psi^s)/2kT)] \tag{F12}$$

The potential ψ^a at the distance $r = a$ from the electrode is accessible from the measurable charge density q^M. It is not possible to make a statement for the distance a without additional assumptions (see solvation of ions).

Now the potential difference $\psi^M - \psi^s$ between metal and solution can be split into two contributions $\psi^M - \psi^a$ and $\psi^a - \psi^s$; ψ^a is the average potential at distance $r = a$. Both the contributions may depend on the charge at the metal, therefore, for the reciprocal differential capacity of the entire double layer follows:

$$\delta(\psi^M-\psi^s)/\delta q^M = \delta(\psi^M-\psi^a)/\delta q^M + \delta(\psi^a-\psi\ )/\delta q^M \tag{F13}$$

and

$$\delta(\psi^M-\psi^s)/\delta q^M = \delta E/\delta q^M = 1/C \tag{F14}$$

for the ideal polarizable electrode at constant temperature and constant composition of the electrolyte. (An ideal polarizable electrode shows no Faraday currents.)

According to eq. (F14) the capacity of the double layer may be formulated as the serial connection of the differential capacity C_i of the "inner layer" (Stern layer or Helmholtz layer) and the differential capacity C_d of the diffuse part of the double layer:

$$1/C = 1/C_i + 1/C_d \tag{F15}$$

Appendix G

Laplace transformation and application on Fick's second law

(This text is for pioneers among the readers)

The Laplace transformation is an often used mathematical tool to solve partial differential equations, such as Fick's second law, in which the concentration is a function of both time and space. They are generally more difficult to solve than total differential equations, where the dependent variable is a function of only one independent variable. An example of a total differential equation of second order is the linearized Poisson–Boltzmann equation: $1/r\, d(r^2\, d\psi/dr)/dr = \kappa^2\, \psi$. It is a second-order derivative.

The order of a differential equation is the order of the highest derivatives, which in the example is a second-order derivative, $d^2\psi/dr^2$. It has been shown in Chapter 2.6.5 that the solution of this equation was easily accomplished.

The solution of Fick's second law: a partial differential equation would proceed smoothly if some mathematical device could be utilized to convert it into the form of a total differential equation. This transformation is based on the operation[72] of Laplace transformation. A brief digression on the nature of this operation is given before using it to solve the partial differential equation involved in the nonsteady-state electrochemical diffusion problems, namely Fick's second law.

Consider a function y of the variable z, that is $y = f(z)$, represented by a plot of y against z. The familiar operation of differentiation performed on the function y consists in finding the slope of the curve representing $y' = f(z)$ for various values of z, that is, the differentiation operation consists in evaluating dy/dz. The operation of integration consists in finding the area under the curve, that is, it consists in evaluating $\int_{z1}^{z2} f(z)dz$.

The Laplace transformation performed on the function $y = f(z)$ consists of two steps:

1. Multiplying $y = f(z)$ by e^{-pz}, where p is a positive quantity which is independent of z
2. Integrating the resulting product ye^{-pz} $(= f(z)e^{-pz})$ with respect to z between the limits $z = 0$ and $z = \infty$

In short the Laplace transform $y = f(z)$ is

$$\int_0^\infty e^{-pz}\, y\, dz \text{ or } \int_0^\infty e^{-pz} f(z)\, dz$$

72 A mathematical operation is a rule for converting one function into another.

Just as one often symbolizes the result of differentiation of y by y', the result of the operation of Laplace transformation performed on y is often represented by a symbol $\bar{y}$. Thus $\bar{y}$ = Laplace transform of y; it is $\int_0^\infty e^{-pz} y dz$.

What happens during Laplace transformation can be easily visualized by choosing a function, say $y = \sin z$. It can be seen that the operation consists in finding the area under the curve ye^{-pz} between the limits $z = 0$ and $z = \infty$. It can also be seen that, apart from having to make the integral converge, the exact value of p is not of significance for, in any case, p disappears after the operation of inverse transformation.

Table G.1: Laplace Transforms of some common Functions*

Function	Transform
A (constant)	A/s
e^{-at}	$1/(s + a)$
sin at	$a/(s^2 + a^2)$
cos at	$s/(s^2 + a^2)$
sinh at	$a/(s^2 - a^2)$
cosh at	$s/(s^2 - a^2)$
t	$1/s^2$
$\mathrm{erfc}[x/2(kt)^{1/2}]$	$e^{-\beta x}/s$

*Adapted from A. Bard, Lr. Faulkner “Electrochemical Methods”. P.659

Application of Laplace transformation on Fick’s second law

It will be shown that Laplace transformation converts the partial differential equation, which is Fick’s second law into a total differential equation, which can be readily solved:

$$dc/dt = D\, d^2c/dx^2 \tag{G1}$$

Since whatever operation carried out on the left-hand side of the equation must be repeated on the right-hand side, therefore, both sides of Fick’s second law will be subject to the operation of Laplace transformation

$$\int_0^\infty e^{-pt}\, \delta c/\delta t\, dt = \int_0^\infty e^{-pt}\, \delta^2 c/\delta x^2 dt \tag{G2}$$

which, by using the symbol for a Laplace transformation, can be written as

$$\bar{\partial}\,\bar{c}/\partial t = D\,\bar{\partial}^2 c/\partial x^2 \tag{G3}$$

To proceed further, one must evaluate the integrals of eq. (G2).

Consider the Laplace transform

$$\bar{\partial}\,\bar{c}/\delta t = \int_0^\infty e^{-pt}\delta c/\delta t\;dt \tag{G4}$$

The integral can be evaluated by the rule of integration by parts as follows:

$$\int_0^\infty e^{-pt}\;\delta c/\delta t\;dt = e^{-pt}\int_0^\infty \partial c - \int_0^\infty\left[\int_0^\infty \delta c\right] de^{-pt} = [e^{-pt}c]_0^\infty + p\int_0^\infty e^{-pt}c\;dt \tag{G5}$$

(Remember $d(u{\cdot}v) = u'v + v'u$.)

Since $\int_0^\infty e^{-pt}c\;dt$ is in fact the Laplace transform of c and, for conciseness, is represented by the symbol $\bar{c}$ since e^{-pt} is zero when $t \to \infty$ and unity when $t = 0$, eq. (G4) reduces to

$$\bar{\partial}\,\bar{c}/\delta t = \int_0^\infty e^{-pt}\;\delta c/\delta t\;dt = -c(t=0) + p\,\bar{c} \tag{G6}$$

where $c\,(t = 0)$ is the value of the concentration c at $t = 0$.

Next, one must evaluate the integral on the right-hand side of eq. (G2), that is

$$D\bar{\partial}^2 c/\partial x^2 = \int_0^\infty e^{-pt}\,D\;\delta^2 c/\delta x^2 dt \tag{G7}$$

Since the integration is with respect to the variable t and the differentiation is with respect to x, their order can be interchanged. Further, one can move the constant D outside the integral sign. Hence, one can write

$$D\;\bar{\partial}^2 c/\partial x^2 = D\;\partial^2/\partial x^2\int_0^\infty e^{-pt}\;cdt \tag{G8}$$

Once again it is clear that $\int_0^\infty e^{-pt}c$ is the Laplace transform, that is, $\bar{c}$ and therefore

$$D\;\bar{\partial}^2\;\bar{c}/\partial x^2 = D\;\partial^2/\partial x^2\;\bar{c} \tag{G9}$$

From the equations

$$\partial c/\partial t = D\;\partial^2 c/\partial x^2$$

$$\bar{\partial}\, \bar{c}/\delta t = \int_0^\infty e^{-pt}\, \delta c/\delta t\, dt = -\, c\,(t=0) + p\,\bar{c}$$

$$D\, \partial^{\bar{2}}\, \bar{c}/\partial x^2 = D\, \partial^2/\partial x^2\, \bar{c}$$

It follows that, after Laplace transformation, Fick's second law takes the form

$$p\,\bar{c} - c\,(t=0) = D\, d^2\, \bar{c}/dx^2 \tag{G10}$$

This, however, is a total differential equation because it contains only the variable x. Thus by using the operation of Laplace transformation Fick's second law has been converted into a more easily solvable differential equation involving $\bar{c}$, the Laplace transform of the concentration.

The initial and boundary conditions for the diffusion process stimulated by a constant current (or flux)

An example

A differential equation results by differentiating an original equation, or primitive, as is called. In the case of Fick's second law, the primitive is the equation that gives the precise nature of the functional dependence of concentration on space and time, that is, the primitive is an elaboration on

$$c = f(x, t)$$

In the process of differentiation, constants are eliminated and since three differentiations, two with respect to x and one with respect to time, are necessary to arrive at Fick's second law. Three constants have been eliminated in the process of going from the precise concentration dependence, which characterizes a particular problem to the general relation between the time and space derivatives of concentration which describes any nonstationary diffusion situation.

The three characteristics, or conditions, as they are called, of a particular diffusion process cannot be rediscovered by a mathematical argument applied to the differential equation. To get at the three conditions, one has to resort to a physical understanding of the diffusion process. Only then one can proceed with the solution of the total differential equation and get the precise functional relationship between concentration, distance and time.

In the following we use a typical electrochemical diffusion problem, which is of general significance, for the discussion.

In an electrochemical system a constant current is switched on at a time which is arbitrarily designated $t = 0$. The current is due to charge-transfer reactions at the electrode–solution interface. At the electrode, species is consumed, which leads to a decreasing concentration of this species at the interface and consequently a concentration gradient for this species is set up. Along this gradient the species under consideration diffuses from the bulk to the electrode. The externally controlled current sets up within the solution a diffusion flux.[73]

The diffusion is described by Fick's second law

$$\partial c/\partial t = D\partial^2 c/\partial x^2 \tag{G11}$$

which reads after Laplace transformation

$$p\,\bar{c} - c(t=0) = D\, d^2\, \bar{c}/dx^2 \tag{G12}$$

Before we can solve the differential equation we must think out the conditions of the diffusion process described earlier, that is, we have to find the three conditions (integration constants).

The nature of one of the conditions becomes clear from the term $c(t = 0)$ in the Laplace-transformed version. It describes the initial condition of the electrolytic solution. Since, before the constant current is switched on and diffusion starts, one has the unperturbed system. The concentration of the diffusing species is the same throughout the system and equal to the bulk concentration c^0:

$$c(t=0) = c^0 \tag{G13}$$

The other two conditions pertain to the situation after the diffusion begins. They are usually known as the boundary conditions, because they refer to what is happening to the boundaries of the system in which diffusion occurs.

Very far from the boundary at which the diffusion sink or source is set up, the concentration of the diffusing species is unperturbed and remains the same as the initial concentration:

$$c(x \to \infty) = c(t=0) = c^0 \tag{G14}$$

Thus the concentration of the diffusing species has the same value c^0 at any x at $t = 0$ and for any $t > 0$ at $x \to \infty$.

The nature of the diffusion flux which is started off at $t = 0$ is characteristic of one particular electrochemical diffusion process and distinguishes it from all others. The characteristic of the diffusion problem under discussion is the switching-on of the constant current. It means that the diffusing species is consumed at a constant rate at the

73 There is a simple relation between diffusion flux and current density; the diffusion flux J is given by the number of moles crossing 1 cm^2 in 1 s. The current density is a charge flux; it is given by the charge zF (z is the valence of the diffusing ion, F faraday constant). The current density i is zF times the diffusion flux J. $i = zFJ$.

interface and the species diffuses across the interface at a constant rate, that is, the flux of the diffusing species at $x = 0$ boundary of the solution is a constant.

It is convenient from any point of view to normalize the diffusion flux. The value of the normalized flux is 1, 1 mol of the diffusing species crossing 1 cm^2 of the electrode–solution interface per second. The unit flux corresponds to a constant current density of 1 amp/cm^2. The normalization of flux scarcely affects the generality of the treatment because the concentration response to an arbitrary flux can be easily obtained from the concentration response to the unit flux.

If one looks to the time variation of the current or the flux across the solution boundary, it is seen that for $t < 0$, $J = 0$, and for $t > 0$ there is a constant flux $J = 1$ corresponding to the constant current switched on at $t = 0$. In other words, the time variation of the flux is like a step. The flux produced in this setup is often known as step function of time. It has to be mentioned that we report here an "ideal switch," in reality there exists a rise time of the step; it depends on the quality of the potentiostat and on the electric properties of the electrode–solution system.

At any instant of time, the constant flux across the boundary is related to the concentration gradient there, through Fick's first law, that is,

$$J_{x=0} = 1 = -D\,(\partial c/\partial x)_{x=0} \tag{G15}$$

The initial and boundary conditions can be summarized:

$$c(t=0) = c^0 \tag{G16}$$

$$c(x \to \infty) = c^0 \tag{G17}$$

$$(\partial c/\partial x)_{x=0} = -1/D \tag{G18}$$

These three conditions describe the special features of the constant (unit) flux diffusion problem. They will now be used to solve Fick's second law.

The concentration response to the constant flux switched on at $t = 0$

The solution of the Laplace-transformed second law of Fick

$$p\bar{c} - c(t=0) = D\, d^2\bar{c}/dx^2 \tag{G19}$$

is facilitated if the second term is zero. This can be attained by introducing a new variable c_1, which is defined by

$$c_1 = c^o - c \tag{G20}$$

where c_1 represents the perturbation of the initial concentration c^0.

The partial differential equation and the initial and boundary conditions must now be restated in terms of the new variable c_1. One obtains

$$\partial c_1/\partial t = D\ \partial c_1/\partial x^2 \tag{G21}$$

$$c_1(t=0)=0 \tag{G22}$$

$$c_1(x\to\infty)=0 \tag{G23}$$

$$(\delta c_1/\delta x)_{x=0} = -1/D \tag{G24}$$

After Laplace transformation of the differential equation becomes

$$p\,\bar{c}_1 - c_1(t=0) = D\ d^2\,\bar{c}_1/dx^2 \tag{G25}$$

Since c_1 $(t = 0) = 0$ it results in

$$d^2\ \bar{c}_1/dx^2 = P\ \bar{c}_1/D \tag{G26}$$

This equation is identical in form to the linearized Poisson–Boltzmann equation and therefore must have the same solution:

$$\bar{c}_1 = A\ \exp\ -(p/D)^{1/2}x + B\ \exp\ (p/D)^{1/2}x \tag{G27}$$

where A and B are arbitrary integration constants to be evaluated by the use of boundary conditions.

The constant B must be zero by virtue of the following argument. From the boundary condition c_1 $(x \to \infty) = 0$, it is clear that after Laplace transformation

$$\bar{c}_1(x\to\infty)=0 \tag{G28}$$

Hence $x \to \infty$, $\bar{c_1} = 0$, but this will be true only if $B = 0$ as otherwise $\bar{c_1}$ will go to infinity instead of zero.

One is therefore left with

$$\bar{c}_1 = A\ \exp\ -(p/D)^{1/2}x \tag{G29}$$

Now one has to find the value of A. Doing this we differentiate the equation with respect to x and obtain

$$d\bar{c}_1/dx = -(p/D)^{1/2}A\ \exp\ -(p/D)^{1/2}x \tag{G30}$$

and with $x = 0$ is

$$(d\bar{c}_1/dx)_{x=0} = -(p/D)^{1/2}A \tag{G31}$$

Another expression for $(d\bar{c_1}/dx)_{x=0}$ can be obtained by applying the operation of Laplace transformation to the constant flux boundary condition $(\delta c_1/\delta x)_{x=0} = -1/D$.

Laplace operation to the left side of the boundary condition leads to $(d\bar{c}_1/dx)_{x=0}$ and the same operation is performed on the right side to $-1/Dp$. Thus from the boundary condition one gets

$$(d\bar{c}_1/dx)_{x=0} = -1/Dp \tag{G32}$$

Hence from eqs. (G31) and (G32) is found

$$A = 1/\ p^{3/2}D^{1/2} \tag{G33}$$

Inserting the expression for A into eq. (G26) follows

$$\bar{c}_1 = \ 1/\ p^{3/2}D^{1/2}\exp - (p/D)^{1/2}x \tag{G34}$$

The last step is to find a way for getting c_1 or c as a function of x and t. One has to find the inverse Laplace transformation, which transfers $\bar{c}$ into c_1 or c, respectively. In other words, one has to find c_1 in the equation

$$\int_0^\infty \exp -p\ t\cdot c_1 dt = \bar{c}_1 = 1/D^{1/2}p^{3/2}\exp\ -(p/D^{3/2})\ x \tag{G35}$$

Since there are extensive tables of functions y and their transform y, it is only necessary to look up the tables in the column of Laplace transforms. It is seen that corresponding to the transform

$$P^{1/2}\exp(-x\ p^{1/2}/D^{1/2}) \tag{G36}$$

is the function

$$2\pi^{-1/2}t^{1/2}\exp(-x^2/4Dt) - x\ D^{-1/2}\mathrm{erfc}(x^2/4Dt)^{1/2} \tag{G37}$$

where erfc is the error function complement defined by

$$\mathrm{erfc}(y) = 1 - \mathrm{erf}(y) \tag{G38}$$

erf(y) is the error function (Fig. 4.21) given by

$$\mathrm{erf}(y) = 2/\pi^{1/2}\int_0^y \exp\ -u^2 du \tag{G39}$$

Hence, the expression for the concentration perturbation c_1 must be

$$C_1 = 1/D^{1/2}[2t^{1/2}/\pi^{172}\exp(-x^2/4Dt\ ) - xD^{1/2}\mathrm{erfc}(-x^2/4Dt\ )^{1/2}] \tag{G40}$$

If one is interested in the true concentration c instead of the deviation c_1 of concentration from c^0, one must use the defining equation for c_1:

$$c_1 = c^0 - c \tag{G41}$$

The result is

$$c = c - 1/D^{1/2}[2t^{1/2}/\pi^{172} \exp(-x^2/4Dt) - xD^{1/2} erfc\ (-x^2/4Dt)^{1/2}] \tag{G42}$$

This is the fundamental equation showing how the concentration of a diffusing species varies with distance x from the electrode–solution interface and with time that has elapsed since a constant unit flux was switched on. In other words, eq. (G42) describes the diffusional response of an electrolytic solution to the stimulus of a flux which is in the form of a step function of time. The nature of response is best appreciated by viewing the concentration profile of the diffusing species that varies as a function of time.

Bibliography

The following books, which have been used for writing this book, are recommended for everybody, who wants to know more about electrochemistry.

Allen Bard, Larry R. Faulkner "Electrochemical Methods", J. Wiley &Sons, New York 1980

Allen Bard, Roger Parsons, Joseph Jordan eds. "Standard Potentials in Aqueous solutions", Marcel Dekker, Inc. New York, Basel 1985

Carl H. Hamann, Wolf Vielstich, "Elektrochemie", Wiley-VCH, Weineheim 2005

E. Budevski, G. Staikov, W.J. Lorenz "Electrochemical Phase Formation and Growth" Verlag Chemie Weinheim 1996

Fritz Beck, "Elektro-organische Chemie, Grundlagen und Anwendungen", Verlag Chemie Weinheim Chemie 1974

Gerd Wedler, Hans-Joachim Freund "Lehrbuch der Physikalischen Chemie" Wiley-VCH Weinheim 2012

Gustav Kortüm, "Lehrbuch der Elektrochemie" Verlag Chemie, Weinheim 1966

Heiz Gerischer, Charles W. Tobias, eds., "Advances in Electrochemical Science and Engineering" Vol. 4, Verlag Chemie Weinheim 1985

Jacek Lipkowski, Philip N. Ross eds. Vol 3: "Frontiers of Electrochemistry", Verlag Chemie, 1994

Vol 1: "Adsorption of Molecules at Metal Electrodes" Verlag Chemie Weinheim 1992

Vol 2: "Structure of Electrified Interfaces" Verlag Chemie Weinheim 1993;

J.O'M. Beckris, A.K.N. Reddy, "Modern Electrochemistry", two volumes, Plenum Press New York 1973

J. O'M. Bockris, B.A. Conway, E. Yeager eds. "Comprehensive treatise of Electrochemistry" 10 volumes; vol.1 The Double Layer, Plenum Press, New York 1980

Klaus J. Vetter," Elektrochemische Kinetik" Springer-Verlag Berlin 1961

P.W. Atkins, "Physical Chemistry", Oxford University Press, Oxford 1982

Roald Hoffmann, "Solids and Surfaces, A Chemist's View of Bonding in Extended Structures" Verlag Chemie Weinheim 1988

Southampton Electrochemistry Group, "Instrumental Methods in Electrochemistry" Ellis Horwood Limited, Chichester 1985

https://doi.org/10.1515/9783110437393-007

List of symbols

a_i	Activity of species *i*
c_i	Concentration of species *i* (mol/L) “molarity”
C_p	Molar heat capacity (J/mol/K) at constant pressure
E	Potential versus a reference electrode (V “Volt“)
E^0	Standard potential (V)
E_e	Equilibrium or “reversibel“ potential (V)
E_{cell}	Difference between cathode and anode potential (V)
F	Faraday (C/mol)
f_i	Activity coefficient (on the basis of concentration or molality or mole fraction)
G	Molar free energy (Gibbs energy) (J/mol)
H	Molar enthalpy (J/mol)
I, i	Current (A)
j	Current density (A/cm^2)
K	Degree Kelvin
K	Thermodynamic equilibrium constant of a reaction
m_i	Moles of species *i* dissolved in 1 kg solvent (mol/kg), “molality“
n	Number of electrons involved in the overall electrode reaction (dimensionless)
n_i	Number concentration of species *i* (cm^{-3} or L^{-1})
p	Pressure, Pascal (N/m^2), 1 atm. = 101325 Pa (N/m^2)
p_i	Partial pressure of species *i*
Q	Electrical charge (C “Coulomb“)
Q^+,Q^-	Electrical charge of cations or anions
q	Charge density (C/cm^3)
R	Gas constant ($J/K^2/mol$)
R	Resistance (Ω)
S	Molar entropy (J/mol/K)
T	Thermodynamic temperature K ("Kelvin“)
t	Celsius temperatue defined by $t = T - T_0$, T_0 = 273.15 K
t	Time (s)
V	Volt (electrical potential difference)
W	Watt
v, v^+, v^-	Stoichiometric factors (integers)
x_A	Mole fraction of species *A*
z, z^+, z^-	Number of electrical charges (integers)

https://doi.org/10.1515/9783110437393-008

Index

https://doi.org/10.1515/9783110437393-009

Bei Fragen zur Produktsicherheit wenden Sie sich bitte an:
If you have any questions regarding product safety,
please contact:

Walter de Gruyter GmbH
Genthiner Straße 13
10785 Berlin
productsafety@degruyterbrill.com